SpringerBriefs in Economics

More information about this series at http://www.springer.com/series/8876

Arup Mitra

Industry-Led Growth

Issues and Facts

 Springer

Arup Mitra
Institute of Economic Growth
Delhi University Enclave
New Delhi
India

ISSN 2191-5504   ISSN 2191-5512   (electronic)
SpringerBriefs in Economics
ISBN 978-981-10-0007-2   ISBN 978-981-10-0009-6   (eBook)
DOI 10.1007/978-981-10-0009-6

Library of Congress Control Number: 2015952767

Springer Singapore Heidelberg New York Dordrecht London

Springer Science+Business Media Singapore Pte Ltd. is part of Springer Science+Business Media
(www.springer.com)

Preface

From the historical experience of the present-day developed nations, we note that one important determinant of economic growth is industrialization. The role of industry is crucial in generating high-productivity employment and enhancing the standard of living of the population. In the process of development, there takes place a structural shift both in the value added growth and in the workforce composition away from the primary sector first towards the secondary and later towards the tertiary sector. This structural change is accompanied not only by a rise in per capita income but also improvement in many other development indicators. It involves upward mobility of individual occupations and incomes and a shift in rural–urban composition of the population. However, in the Indian context, the share of manufacturing in the total workforce has been dwindling at a low level of 11 % or so even after experiencing rapid economic growth in the last several years.

This study focuses on the manufacturing sector in the Indian context. Though both the components—organized and unorganized—have been looked into, the emphasis is laid on the employment aspect of the organized manufacturing group so as to identify the subsectors which have the potentiality to grow and generate productive employment opportunities. Since the unorganized segment is often characterized by low productivity, generating meagre earnings, the policy focus has to be on reducing the vulnerability of these units. The organization of the study is as follows. In Sect. 2, we briefly review some of the existing studies on manufacturing employment. Section 3 presents an empirical analysis of the organized manufacturing at a fairly detailed level of three-digit groups of industries. It analyses growth, employment, productivity and capital–labour ratio across various industry groups. Remunerations to workers and contractualisation process are discussed in the backdrop of the labour market regulation issues. Section 4 focuses on the unorganized manufacturing component. The inter-industry linkages are considered in Sect. 5 through the input–output framework. Section 6 refers to issues relating to skill shortage and the poor employability of the workforce. Based on the firm-level data, Sect. 7 examines the issues related to innovation and employment. Finally, Sect. 8 summarizes the major findings.

   This work draws heavily from the ILO Asia-Pacific Working Paper (Can Industry be the Key to Pro-poor Growth? An Exploratory Analysis for India, December 2013). The author would like to thank Dr. Sher Verick, Dr. Paul Comyn and other anonymous reviewers for their detailed comments and suggestions, which helped enrich the analysis. Thanks are due to Professor B.N. Goldar for discussing some of the issues raised in the text. The chapter on innovation and employment draws from a larger study pursued under the IDRC-TTI grant, Institute of Economic Growth (IEG), Delhi. The author is grateful to the library and computer unit staff, IEG and the publishing house, Springer, for their support and cooperation.

Arup Mitra

Contents

About the Author

Arup Mitra is a professor of economics at the Institute of Economic Growth, New Delhi, India. His research interest includes issues in the area of development economics, urban development, labour and welfare, industrial productivity growth and employment, services sector growth and trade in services, and gender studies. He has published in a number of international and national journals. He worked as a senior researcher at ILO (Geneva), was offered visiting fellowship at the Institute of Developing Economies (Tokyo) and held the Indian Economy Chair at Sciences Po. (Paris). The Indian Econometric Society offered him the Mahalanobis Memorial Gold Medal for his outstanding contribution in the field of quantitative economics. His work has been cited in the *Handbook of Regional and Urban Economics* (Elsevier, 2004), and he has also contributed in the *Encyclopaedia of Life Support Systems* (EOLSS), *Mathematical Models in Economics* (Ed. Zhang, W.), developed under the auspices of the UNESCO, and *Encyclopaedia of Sustainability*, Great Barrington, MA: Berkshire Publishing.

About the Book

The book explores, for India and other developing countries, the potential role the organized manufacturing sector could play as an engine of growth. Alongside growth, can this sector generate adequate employment opportunities to facilitate the transfer of labour from the agriculture sector? The book identifies the major constraints that result in limited demand for labour in the organized manufacturing sector. Beyond technological aspects, skill shortage is an important factor, resulting in sluggish labour absorption. Further, the labour market laws are not necessarily the root cause of sluggish employment growth in the organized manufacturing sector. The development of technologies that are appropriate for labour-surplus countries like India is instrumental to employment creation. Though innovation is generally assumed to be capital-intensive in nature, the book argues that innovation nevertheless has a positive effect on employment in absolute terms. Lastly, the main policy issues are highlighted in terms of the priority that should be assigned to industries which can contribute to employment growth and skill formation for improving the employability of the available labour force, and to which innovations should be pursued, with a specific focus on pro-poor growth objectives.

Abstract

The gross value added growth rate continued to be a little above 9 % per annum during 1998–1999 to 2007–2008. However, the employment growth rate, compared to the 1990s (prior to 1998–99), decelerated marginally to 2.6 % per annum, more so in the case of employees other than workers. Labour productivity defined as the value added per person employed grew at almost 7 % per annum. Wages per worker remained almost stagnant while the remuneration per person shot up significantly, implying a substantial growth in the salaries per employee (excluding workers). Industries which dominated in terms of employment share did not necessarily unravel a fast employment growth. There is a strong positive correlation between the average value added growth and total employment growth measured across all the three-digit manufacturing groups, implying growth is essential for employment generation.

For the aggregate manufacturing sector, the elasticity of total employment with respect to value added is 0.43. Industries which recorded an employment elasticity of up to 0.4 are many, though only handful of them had an employment elasticity of more than 0.55. These industries may be targeted for providing a boost to employment growth; but given their modest share, it is unlikely that they will succeed in raising the overall employment growth of the manufacturing sector in a significant manner as specified in the National Manufacturing Policy (NMP).

Since wages do not constitute a large component of the total cost, a more careful analysis has to be pursued to identify what restricts industrial expansion and employment creation rather than simply blaming the labour laws. Mechanisms to initiate flexibility are already in motion even without labour market deregulation being carried out formally. Labour laws alone cannot be held responsible for sluggish employment growth: the unorganized manufacturing to which the labour laws are not applicable has been shrinking in relation to employment generation.

As regards inter-industry linkages, we may conclude that industrial deceleration in the heavy goods sector can reduce the input demanded from the labour as well as capital-intensive sub-sectors. Thus, the growth and employment in the labour-intensive sub-sectors may suffer which in turn may affect adversely the overall employment growth in the manufacturing sector. This may also have spillover (negative) effects on the rest of the economy and the pace of employment generation and the effectiveness of the industrial sector in reducing poverty.

The index representing the difference between the skill level of the population and the workers is estimated at 73.11. In other words, the difference between the skill levels of the potential labour supply and those already working is sizeable. The intertemporal skill mismatch index as estimated from the distribution of workers in each of the activities across various skill levels is again substantial, indicating that over time jobs in manufacturing are becoming more skill-biased. In the earnings function higher returns to education are evident. Thus, those who are able to acquire skill are able to secure employment with higher levels of earnings as demand exists for the skilled variety of labour due to rise in skill intensity of the technology in the manufacturing sector.

Innovation is also seen to contribute to employment creation though it is usually feared that innovation leads to capital-intensive methods of production, reducing employment sizably. With a focus on pro-poor growth issues if innovation is undertaken domestically, employment growth is certain to take place along with output growth.

Finally, the policy issues are highlighted in terms of (a) priority to be given to industries which can contribute to employment growth, (b) skill formation for improving the employability of the available labour force and (c) innovation is important for pro-poor growth to take place.

Keywords Industry · Growth · Employment elasticity · Wage · Labour productivity · Unorganized · Inter-industry linkage · Skill · Innovation

Industry-Led Growth: Issues and Facts

1 Introduction

India emancipated from the prolonged phase of conservative Hindu rate of growth almost a quarter century back. However, given its vast labour supplies the labour demand, particularly for those who are less skilled and poorly endowed with human capital, has been growing sluggishly. To be specific, whether economic growth is able to generate employment opportunities on a large scale, particularly for the unskilled, semi-skilled and the less-educated labour force, is an important question that has been bothering the development economists since long. Second, what impact the economic reforms in various sectors have made on the employment front is another important aspect that needs a thorough investigation. Accordingly, the future strategy has to be developed and sectors and policies that hold potentiality and prospects for employment growth need to be identified. After all, for economic growth to be inclusive or pro-poor, productive employment generation has to take place in a significant way.

It has been widely noted that growth alone is not sufficient to bring in any major improvement in economic and social wellbeing, particularly of those who are located at the lower echelons of the socio-economic ladder. Rapid growth in productive employment opportunities can distribute the benefits of growth among the deprived lot. In other words, employment growth at wages higher than the minimum subsistence level of consumption is crucial for poverty reduction and also to create a stable society that would be free from social turmoil and insurgency.

From the historical experience of the present day developed nations, we note that one important determinant of economic growth is industrialization. The role of industry is crucial in generating high-productivity employment and enhancing the standard of living of the population. In the process of development, there takes place a structural shift both in the value added and work force composition away from the primary sector first towards the secondary and later towards the tertiary sector. This structural change is accompanied not only by a rise in per capita

© The Author(s) 2016
A. Mitra, *Industry-Led Growth*, SpringerBriefs in Economics,
DOI 10.1007/978-981-10-0009-6_1

income but also improvement in many other development indicators. It involves upward mobility of individual occupations and incomes and a shift in rural-urban composition of the population (Kuznets 1966). However, in the Indian context, the share of manufacturing in the total work force has been dwindling at a low level of 11 % or so even after experiencing rapid economic growth in last several years.

Szirmai and Verspagen (2011), in the context of developing countries, point out that manufacturing since 1990 is becoming a more difficult route to growth than before. They also find interesting interaction effects of manufacturing with education and income gaps. Dellas and Koubi (2001) argue that the industrialization of labour is the main engine of growth during the early stages of economic development. Often, equipment investment has played a less important role than non-equipment investment. Besides, the outcome is growth enhancing when it either encountered a substantial industrial labour force or fostered a large increase in the share of industrial employment. These findings, as Dellas and Koubi (2001) view, draw attention to the effects of investment on the composition of the labour force; and unlike recent claims emphasizing industrialization via equipment investment, they suggest that employment industrialization policies may hold the key to success in the developing countries.

The most important factor which aggravates the mismatch between the demand for and supply of labour is the sluggish employment growth in the high-productivity industrial sector. This could be due to the limited spread of the industry and/or adoption of capital-intensive technology, leading to a residual absorption of labour in the low-productivity informal sector with meagre earnings accruing to the workers and compelling them to reside in slums and squatter settlements. Globalization has compelled countries to enhance growth. Several growth-oriented strategies, that include trade openness, FDI-inflows and capital mobility, including technology transfer, have been adopted in a big way. The argument, which is usually given in favour of technology transfer, is that the wheel that has already been adopted need not have to be rediscovered if countries seek to be cost efficient.[1] However, one important hypothesis in the context of sluggish employment growth in the industrial sector relates to the acquisition of capital-intensive technology imported from abroad.[2] The import of new technology, which is primarily capital intensive and skill intensive, results in increased demand for skilled workers and not for the less-skilled ones (Wood 1997).[3]

---

[1]It is argued that countries further from the frontier have lower R&D returns, implying that the cost of innovation is more in a poor country than in a rich country. Hence, it is still cheaper for a latecomer to buy the technology already invented by others than to re-invent the wheel though it is widely noted that international technology does not come cheap (UNIDO 2005).

[2]Research for various Latin American countries is indicative of widening impact of trade on wage inequality, and more importantly this is spearheaded by the notion of skill-biased technological change induced through trade (Hasan 2003).

[3]Hasan and Mitra (2003) noted that trade is enhancing rapidly the premium to skills in developing countries due to the skill-biased production technologies embodied in imported inputs. Globalization raises capital flows from developed to developing countries which means that, even

Another line of argument is based on labour market regulations. The policy circle usually believes that labour laws in the Indian context are extremely outdated and pro-employee which in turn tends to reduce labour absorption. Due to the lack of labour market flexibility, economic growth and employment generation, as it is believed, cannot receive an impetus in a sustainable manner. The advocates of economic reforms lay considerable emphasis on labour market deregulations. This is because globalization and shifts in the production activities are expected to impact on the labour market outcomes such as wages and labour productivity. Second, and more importantly, for other reforms, in the area of trade for example, to be successful, labour market reforms are considered as essential prerequisites (Hasan 2003). Scholars generally view that the labour markets in developing countries are rigid in terms of work practices, wages, hiring and firing policies, etc., and all this has been attributed to the existing labour laws (Fallon and Lucas 1991).[4] However, there are many other factors affecting investment, most notably, access to land, infrastructure, skilled workforce, etc.

A third line of argument, particularly from the supply side, refers to skill shortage, pointing to the poor employability of the vast sections of the labour force. The lack of skill forces many to get residually absorbed in low-productivity activities.[5]

This study focuses on the manufacturing sector in the Indian context. Though both the components—organized and unorganized[6]—have been looked into, the emphasis is laid on the employment aspect of the organized manufacturing group so as to identify the sub-sectors which have the potentiality to grow and generate productive employment opportunities. Since the unorganized segment is often characterized by low productivity and generating meagre earnings, the policy focus has to be on reducing the vulnerability of these units. The organization of the study is as follows.

(Footnote 3 continued)

without technology imports, capital output ratios in developing countries would rise and, given the complementary relationship between capital and skills, this would raise the relative demand for skilled labour (Mayer 2000). Johanson (2004) argues that it is not possible to disentangle the interlinked factors responsible for changes in skill demands, but at least three main forces are at work in increasing the demand for skills worldwide which are technological change, changes in work organization and trade openness. Rodrik (1997) further argued that trade, while generating more employment opportunities, may also diminish the bargaining power of workers, thus resulting in deterioration of working conditions.

[4]The World Bank report on "Doing Business" in 2005 estimated that India is ranked at 48th in terms of 'Rigidity in Employment Index' compared to China's rank of 30.

[5]See Times of India, April 22, 2013.

[6]Organized manufacturing sector includes units engaging more than 20 workers irrespective of use of power and units with 10–19 workers using power. The size of the organized segment in the manufacturing sector is around 60 % in terms of value added. We do not report the relative size in terms of employment because the definition of worker in the organized sector as given by the Annual Survey of Industries and that given by the National Sample Survey Organisation (NSSO) for the unorganized manufacturing are not comparable. NSSO uses a very liberation definition of worker so as to include those who have been attached to the unit.

In the next chapter, we briefly review some of the existing studies on manufacturing employment. Section 3 presents an empirical analysis of the organized manufacturing at a fairly detailed level of three-digit groups of industries. It analyzes growth, employment, productivity and capital–labour ratio across various industry groups. Remunerations to workers and contractualization process are discussed in the backdrop of the labour market regulation issues. Section 4 focuses on the unorganized manufacturing component. The inter-industry linkages are considered in Sect. 5 through the input-output framework. Section 6 refers to issues relating to skill shortage and the poor employability of the work force. Based on the firm-level data, Sect. 7 examines the issues related to innovation and employment. Finally Sect. 8 summarizes the major findings. The database of the study is drawn from four major sources: Annual Survey of Industries (ASI) on the organized manufacturing sector, National Sample Survey (NSS) data on unorganized manufacturing, the NSS data on employment–unemployment in order to throw light on skill issues and firm-level data compiled by ACEEQUITY.

2   Review of Select Studies

Industry's role as the engine of growth has been highlighted in the literature primarily because industry is expected to sustain its growth and also contribute to the growth of the other sectors, thus raising the overall growth of the economy. Kaldor (1967) rationalized the structural transformation away from agriculture towards industry mainly in terms of productivity growth in the latter. Several others have, however, argued in the recent years that services can provide an alternative to manufacturing as a driver of economic growth based on India's development path (i.e. rapid growth of IT and business processing outsourcing). Dasgupta and Singh (2005) argue that the rapid expansion of the services sector in India has had an impact on the manufacturing sector as well.

In the context of employment generation, the displaced labour from the agriculture sector without much skill is expected to get absorbed in the industrial sector as there can be huge labour demand resulting from rapid industrialization. More importantly, there may not be a sizeable mismatch in terms of quality between labour available and labour required in the industrial sector though it can be a serious problem for the services sector. However, the observed outcomes have not been in line with what was supposed to result in. In a large number of developing countries, the relative size of the industrial sector both in terms of GDP and more so in terms of workforce, has remained negligible. Even if its income share has risen to some extent, the employment generating capacity of the industrial sector has remained insignificant because of the adoption of capital-intensive technology. Below, we present a detailed review of some of the studies for India and other countries.

Kannan and Ravindran (2009) analyzed the organized industrial employment from 1981–1982 to 2004–2005 in India. For the period as a whole as well as for two separate periods—the pre- and post-reform phases—the picture that emerges is termed by the authors as one of "jobless growth". Due to the combined effect of two trends which have canceled each other out, no employment gain is perceived. One set of industries was characterized by employment-creating growth while another set by employment-displacing growth. Also, over this period, there has been acceleration in capital intensification at the expense of employment creation. A significant part of the resultant increase in labour productivity was retained by the employers as the product wage did not increase in proportion to output growth. The workers as a class thus lost in terms of both additional employment and real wages. The study by Das and Kalita (2009a, b) makes an attempt to address the issue of declining labour intensity in India's organized manufacturing, particularly in labour-intensive sectors. Using primary survey data covering 252 labour-intensive manufacturing–exporting firms across five sectors—apparel, leather, gems and jewellery, sports and bicycles—for the year 2005–2006, the authors discern the factors which constrain the generation of employment in labour-intensive firms. Several constraints that arise in the path of employment generation in the labour-intensive sectors include non-availability of trained skilled workers, infrastructure bottlenecks, low levels of investment, non-competitive export orientation resulting in infrastructural bottlenecks and rigid labour rules and regulations.

Hijzen and Swaim (2007) looked at the implications of off-shoring for industrial employment for 17 high-income OECD countries. Their findings indicate that off-shoring has no effect or a slight positive effect on sectoral employment. Off-shoring within the same industry ("intra-industry off-shoring") reduces the labour intensity of production, but does not affect the overall industrial employment. On the other hand, although inter-industry off-shoring does not affect labour intensity, it may have a positive effect on overall industrial employment. These findings suggest that the productivity gains from off-shoring are sufficiently large. Csacuberta et al. (2004) analyzed the impact of trade liberalization on labour and capital gross flows and productivity in Uruguayan manufacturing sector. Though higher international exposure implied a higher job creation, unionization dampened the process. Industry concentration mitigated the destruction of jobs but could not actually contribute to the job creation process sizably.[7]

In relation to the import of technology from developed countries, mismatches are said to exist between the technological requirements of the developing countries and the available technology (Pack and Todaro 1969). Technology innovated in the developed world is mainly labour-saving and skill-intensive as it has to suit the situation prevailing in the developed economies. On the other hand, developing countries are mainly labour surplus and skill-scarce economies and hence, the objective of employment growth along with economic growth gets defeated when technology is largely imported from the western world. As Azeez (2006) points out distinctly, a new technology gets embodied in capital goods, and therefore, import of capital goods is often considered as import of technology. Once imported capital

---

[7]Given the stagnating industrial experience of the Latin America Brady et al. (2011) based on the dataset for the period 1986 through 2006 observed the negative effects of various factor.

good is put into operation, the technological progress realized in the country of origin will be incorporated into the production process (UNIDO 2005). Hence, it is still cheaper for a latecomer to buy the technology already invented by others than to re-invent the wheel though it is widely noted that international technology does not come cheap (UNIDO 2005). Thus the debate relating to the nature of technology gets shifted to the import of capital goods. Though the import of technology from developed countries may be expected to enhance the productivity in developing countries, there are plausible doubts about this effect. Chakravarty (1987) noted that with imports of capital goods on a significant scale, domestic costs of production are unlikely to come down since developing countries might be importing expensive capital goods. Chandrasekhar (1992) argued further that in the Indian context imports of capital goods have acted as substitutes for domestic production of capital goods, imposing a social cost in the form of unutilized capacity. And this made the domestic firms operate at high unit cost of production.

On the other hand, Hasan (2003) presents evidence from panel data on Indian manufacturing firms, suggesting that imported technologies have a significant effect on productivity. The point regarding technology is crucial, I believe. Technologies (and management systems) are increasingly determined globally. Therefore, it is not surprising that manufacturing in India remains more capital-intensive as the newest assembly processes, etc. rely heavily on technology and skills. This has important policy implications: the large scale, low-cost labour-intensive manufacturing is unlikely to emerge in India because, to be competitive, India has to use these technologies (given the level of input costs in the country). Using data for 33 Indian manufacturing industries in India for the period from 1992 through 2001, Pandit and Siddharthan (2006) show that technology imports, through joint ventures and MNE participation, influence employment positively. They noted that employment growth, production of differentiated products, skill intensity of the work force and technological upgradation go hand in hand.

Another line of argument asserts that the adoption and adaptation of these international technologies are indeed costly because of tacit knowledge and circumstantial sensitivity of technology (Evenson and Westphal 1995). Unless an importing country has significant technological capability, it cannot fully utilize the imported technology. Besides, imported technology may require more skilled than unskilled workers while developing countries are usually abundant in the latter. Acemoglu and Zilibotti (2001) argue that due to the difference in skill scarcity, technology in developed countries tends to be skill-intensive and is inappropriate for developing countries. Thus the potential productivity of imported technology cannot be realized in developing countries. Berman and Machin (2000) empirically showed the skill-bias of technological change especially in middle-income countries. Mitra (2003) and Kato and Mitra (2008) presented evidence from developing countries and India, respectively, to suggest that the possibility to import technology reduces the labour intensity in the production process. Further, imported technology also tends to reduce the technical efficiency as the new technology cannot be operated optimally.

If the import of technology enhances productivity as well as promotes employment, the choice is most desirable. Such a possibility, though empirically unprecedented, exists at least theoretically: for example, a new technology, if accompanied by technological progress, can bring in upward shift in the production frontier, which would mean higher levels of output for the given levels of inputs: capital productivity and labour productivity both may shoot up. Even if the new technology becomes more labour intensive, the rise in value added can still be more than the rise in employment, and hence, labour productivity can actually shoot up along with an increase in labour demand. Further, if we presume expansion in the scale of production, which could be due to the wider applicability of the new technology, employment increase is assured. Even when labour productivity declines in response to labour-intensive technological progress, it may still be desirable to pursue as the new configuration of capital-labour may entail substantial decline in the cost of production. This is more likely to happen, particularly given the modest remuneration for labour in the developing countries in comparison to the international standards. However, the developed countries being mostly characterized by labour shortage, the new technology innovated by them are labour-saving type, which can contribute to growth without generating employment opportunities.

As Das and Kalita (2009a, b) reviewed, the period of the 1980s is often called the decade of 'jobless growth' in Indian manufacturing, since the revival in output growth during this period was not accompanied by the adequate generation of employment. Only 484,000 new jobs were generated in India's registered factory sector or organized sector between 1979–1980 and 1990–1991 (Thomas 2002). One of the explanations put forward was that of difficulty in labour retrenchment after the introduction of job security regulations in the late 1970s, which forced employers to adopt capital-intensive production techniques (Fallon and Lucas 1991). According to another view, the slowdown in employment growth resulted from a strategy of capital deepening pursued by the firms, an important reason for which was the increase in the real cost of labour in the 1980s in India (Ghose 1994). A study undertaken by the World bank (1989) also asserted that the sharp deceleration in the employment growth in the factory sector in the 1980s could be explained by the acceleration in product wages, attributed to union-push. Papola (1994), Bhalotra (1998), and Nagaraj (1994), have highlighted that during the 1980s, there was faster growth of industries with low labour intensity and slower growth of industries with high labour intensity in India and indicated a more intensive use of the workforce in the 1980s, resulting in the slowdown of employment growth.

During the period 1990–1991 to 1997–1998 Nagaraj (2000) pointed out that faster employment generation in the organized manufacturing sector in India, as highlighted by Goldar (2000) and Thomas (2002), was due to the investment boom during that decade. In his later study, Nagaraj (2004) argued that faster employment generation in the organized manufacturing sector was restricted mainly to the first half of the 1990s. As the boom went bust, there was a steep fall in employment during the second half of the 1990s. The relative cost of labour did not seem to matter in employment decisions, as the wage–rental ratio declined steadily (Das and Kalita 2009a, b).

Goldar (2011a, b) argues that employment in India's organized manufacturing sector increased in recent years at the very rapid rate of 7.5 % per annum between 2003–2004 and 2008–2009. He further rationalizes it in terms of labour market deregulation, contributing to manufacturing employment growth. The value added growth also picked up to a double-digit figure during the same period (13.7 % per annum). However, as Nagaraj (2007) argued, the fine print of exemptions and loopholes built into the labour laws provide sufficient flexibilities to the industrial firms and hence, labour regulations could not be the cause of deceleration in employment growth in the past. By the same logic labour market deregulation, therefore, could not be treated responsible for rapid employment growth in the recent period. He argues that the recent manufacturing employment boom could be merely a recovery of employment lost over the previous nine years. He also points out that the correlation coefficient between employment elasticity and labour reforms index across states is not statistically significant (Nagaraj 2011). Another way of rationalizing this employment boom could be in terms of change in the regional distribution of industrial employment growth. Some of the states which have not been industrialized registered a rapid employment growth rate during the recent period. The states like Chattisgarh, Haryana, Punjab, Goa, Jammu and Kashmir, Himachal Pradesh, Orissa and Uttarakhand witnessed a double-digit employment growth whereas most of them (except Goa) had experienced either a sluggish or negative employment growth in the earlier years (1998–1999 through 2003–2004).

However, as a response to Nagaraj (2011), Goldar (2011b) observed that first of all the employment growth in India during 2003–2004 through 2008–2009 has been accompanied by a sharp rise in value added too. This sort of a steep rise in value added was also experienced during the eighties (1980–81 to 1989–90) and nineties (1992–93 to 1996–97) but those phases of growth were accompanied by a negligible employment growth. The rise in employment growth in organized manufacturing in the recent years is not due to a shift in the industrial structure in favour of the labour-intensive industries, as the structure has remained by and large constant. On the other hand, the share of private limited companies which reveal higher employment intensity than the public limited companies or factories classified as individual proprietorships has increased during this phase and has possibly con-tributed partly to this rapid employment growth. On the other hand, Goldar (2011a, b) also notes the contribution of the labour market reforms to employment growth. Though in the Indian context labour market reforms have not been carried out explicitly, informal channels have been followed for its implementation. The index of labour market reforms calculated across states tends to explain the differences in the employment growth rates across states.

With this background, we now turn to the empirical analysis of the three-digit industry groups in the organized manufacturing sector. The period under consid-eration is 1998–1999 through 2007–2008. Since the NIC 1998/2004 is very dif-ferent from the earlier one, involving comparability problem relating to various industry groups, the past data has not been considered in the present study. However, based on our previous study (Mitra and Bhanumurthy 2006), we present a snapshot of the earlier phase ending at 1997–1998.

3 Empirical Analysis: Organized Manufacturing Sector

3.1 Performance Till 1997–1998

The performance of organized manufacturing in India in terms of the growth rate in gross value added showed marked improvement in the 1990s compared to the earlier period (Table 1). Whether this growth had also resulted in faster employment elasticity or not, has been a matter of serious concern. In terms of mere growth rates of course, both the number of workers and total persons increased from a mere 1 % per annum during the deregulated regime (1984–85 to 1990–91) to around 3 % per annum over the 1990s though this growth has been only marginally above the growth rate that was experienced during the regulated regime (1973–1974 through 1984–1985). Man-days per worker and man-days per person grew negligibly during the 1980s and 1990s. For employees other than workers, it is not a matter of serious concern because they are full-timers. But for 'workers' category, man-days per worker is an important determinant of earnings, and hence the stagnancy in man-days per worker may have serious implications in terms of workers' income as it may have resulted from the decline in full-time jobs to the workers in the organized industrial sector. However, the constancy of man-days per worker or person may also have resulted from a rise in outsourcing and subcontracting and assignment of jobs on piece-rate basis. Also, it could be an outcome of exhaustion of scope to utilize labour more intensively (Bhalotra 1998 and Nagaraj 1994). For example, the contract workers were already utilized to the optimum and there was

Table 1 Growth rate of select variables (percent per annum)

Variables	1973–1974 to 1984–1985	1984–1985 to 1990–1991	1990–1991 to 1997–1998
Gross value added	6.4	7.9	9.4
Gross output	7.6	8.4	8.6
No. of workers	2.8	1.1	3.1
Man-days per worker	1.9	0.2	0.2
No. of persons employed	2.9	1.1	3.2
Man-days per person employed	1.7	0.3	0.2
Wages per worker	3.0	3.2	2.7
Emoluments per person Employed	2.4	2.9	3.3
Fixed capital	7.1	6.4	10.8

Note
1. Gross output and value added have been deflated by the wholesale price index of the corresponding product group, and fixed capital, by the combined price index of machinery and metal products with 1981–1982 as base
2. Persons include workers and other employees inclusive of administrative and managerial staff
Source Annual Survey of Industry Data (compiled by Economic and Political Weekly Research Foundation)

hardly any scope for further increase in the man-days per worker. Since the scope to utilize labour more intensively was possibly exhausted, firms were forced to employ additional workers in the 1990s, reflected in higher employment growth rate.

The increase in the employment growth rate in the organized manufacturing in the 1990s, particularly between 1990–1991 and 1995–1996, could also be explained by the huge expansion that took place in the early reform period. Both domestic and foreign investors invested at large quantities during this period with over-expectations about the future prospects demand in the Indian economy and led to expansion in the capacity. This possibly led to an increase in the employment growth rate in organized manufacturing, particularly in the private and joint sector. But, as output started declining or stagnating in the late 1990s, this resulted in capacity underutilization, which might have resulted in job losses (Nagaraj 2004).[8] Despite this downturn, some argue that the employment growth in organized manufacturing has increased in the 1990s compared to the 1980s (Goldar 2000).

Wages per worker shows a fall in the growth rate, marginal though, during the 1990s. (This fall in growth of wages may also be one of the reasons for increase in the employment growth in the nineties[9]). However, emoluments per person did not reveal so (Table 1). Quite clearly, the earnings of the skilled/educated employees other than the workers seem to have increased faster than those of the workers over 1990–1991 through 1997–1998.[10]

For the period 1973–1974 through 1997–1998, the employment elasticity in the organized manufacturing was seen to be quite different from what was noted in other studies.[11] However, the overall level of employment elasticity (measured particularly in relation to workers and not employees) was on the low side considering the fact that Indian economy did not yet appear to have crossed the so-called Lewis turning point in the labour market. On the other hand, although the employment elasticity with respect to value-added declined at the aggregate level in the reform period, the extent of decline was nominal. This picture at the disaggregated level of industry groups is quite different though: while around one-fifth of the industries at the three-digit level experienced rising elasticity, in the rest it remained either very low or stagnant and the reforms have had no impact on the elasticity. However, there are industries which are characterized by high labour intensity and they also have experienced reasonably high employment elasticity (0.5 or more) and fast growth in value added for a period of almost three decades: country liquor (223), tobacco (226), bleaching of silk (246), bleaching of jute and mesta (257), textile garments (265), footwear (291), wearing apparel (292), leather (293), man-made fibres (308), bicycles (376), bullock carts (378), and jewellery articles (383). If we keep

---

[8]Nagaraj (2004) argues that in the second half of 1990s, organized manufacturing sector has lost 15 % of workers across the states and industry groups, mostly due to VRS in public sector and retrenchments and lay-offs in the private sector followed by relaxed labour laws in the country.

[9]Goldar (2000).

[10]Reforms were initiated in July 1991 in India.

[11]Majumdar and Sarkar (2004) find that employment elasticity has dropped from 0.99 in 1974–80 to 0.33 in 1986–96, although in the intermittent period (i.e., between 1980–86) it was–0.16.

aside the criterion of employment elasticity and consider industries which are highly labour intensive and have shown fast growth rates in value added over the decades, the list is as follows: grain milling (204), bicycles (376), country liquor (223), bidi (226), textile garments (265), footwear (291), wearing apparel (292), weaving and finishing of cotton khadi (232), weaving and finishing of cotton textiles on power-loom (234), preparatory operations (253), manufacture of wearing apparel of leather and substitutes of leather (292), manufacture of consumer goods of leather and substitutes of leather other than apparel and footwear (293), manufacture of bullock carts, push carts and hand carts (378), manufacture of jewellery and related articles (383), manufacture of sports and athletic goods (385). This list has considerable overlaps with the former. These industry groups are important indeed from policy point of view as they seem to meet the objectives of pro-poor growth. Another conclusion that emerges from this analysis is that the theoretical wage-employment relation is not very important in the manufacturing sector. This could be due to the existing institutional mechanisms in wage fixation.

3.2 Findings Over 1998–1999 Through 2007–2008

Since economic reforms in the Indian context were initiated in 1991, it would have been reasonable to consider the entire period starting from 1990–1991 till date in our empirical analysis. However, till 1997–1998, the ASI data followed the National Industrial Classification (NIC)-1987 which are not pretty comparable with the classification followed thereafter. Since the NIC-1998 and NIC-2004 are by and large the same, a comparable series of various three-digit groups could be generated for the period 1998–1999 through 2007–2008. However, after 2007–2008 we could not consider the figures because of the comparability problem again, arising from the latest NIC-2008.

In our dataset, the nominal variables have been converted into real terms: the value added figures have been deflated by the WPI (1993–94 base) of the closest commodity group, the detail of which is provided in Table A.1 in the appendix. Fixed Asset is deflated by the price index of machinery and machine, and the real wages and salaries are derived on the basis of consumer price index for industrial workers (1993 base).

Gross value added growth rate continued to be a little above 9 % per annum during 1998–1999 to 2007–2008 (Table 2). However, the employment growth rate which was already low during 1990–1991 to 1997–1998 decelerated marginally, more so in the case of employees other than workers.[12] As a result, labour productivity defined as the value added per person employed grew at almost 7 % per annum.

---

[12]The employment growth rate in the organized manufacturing sector over 1998–99 through 2007–08 as per the ASI data is however higher than the total employment growth rate shown by the NSS employment-unemployment survey over 2004–05 to 2009–10 though the ASI growth rate is quite close to the NSS estimate over 1999–2000 to 2004–05.

Table 2 Growth Rate of Select Variables (percent per annum)

Variables	Rate of growth (% p.a.)
Gross value added	9.45
No. of workers	2.98
No. of persons employed	2.58
Wages per worker	0.20
Emoluments per person employed	5.31
Fixed capital	4.34
Labour productivity (value added per person employed)	6.87
Capital-labour ratio (fixed capital per person)	1.75

Source Annual Survey of Industry Data (compiled by Economic and Political Weekly Research Foundation)

Wages per worker remained almost stagnant while the remuneration per person shot up significantly, implying a substantial growth in the salaries per employee (excluding workers). It is again the fixed capital the growth rate of which decelerated to almost half, resulting in a sluggish growth in capital-labour ratio. Since Indian entrepreneurs had already accumulated a huge stock of capital evident from a sustained growth in fixed capital over the preceding period (Table 1), the decline in the growth rate over the recent phase (from 1997–1998 to 2007–2008) does not come as a surprise.

The distribution of value added and employment across various three-digit industry groups seems to be evenly spread out with a few exceptions (Table A.1 in the appendix). For example, only each of the following industry groups—153, 154, 160, 171, 181, 241, 242, 269, 271 and 291—accounted on an average for nearly 3 or more percentage of the total manufacturing employment. Many of these industries and a few more (232, 292 and 341) also accounted for a 3 or more percentage share in value added terms. Interestingly, not too many of them recorded a rapid employment growth rate of 4 or more percent per annum over the period 1998–1999 to 2007–2008 (i.e. 181, 232 and 269). As an opening remark, industries which dominated in terms of employment size did not necessarily unravel a fast employment growth.

Over this phase, several industries grew rapidly in terms of value added (Table 3). The poorly performing industries have been, in fact, very few in number.[13] In addition, another group of six industries reported a positive but less than 4 % per annum growth rate in value added.[14] Total employment growth in all these industries

---

[13]The worst performing industries with a negative value added growth are about 243 (manufacture of man-made fibres), 313 (manufacture of insulated wire and cables), 333 (manufacture of watches and clocks) and 241(manufacture of basic chemicals) which registered a negative value added growth rate.

[14]151 (production of meat, fish, fruit, vegetables etc.), 152 (manufacture of dairy product), 251 (manufacture of rubber products), 160 (manufacture of tobacco products), and 154 (manufacture of other food products) and 353 (manufacture of aircraft and spacecraft).

Table 3 Poorly performing industries (negative value added growth or positive but less than 4 % p.a.)

Ind. Code	Workers	Employees other than workers	Total persons engaged	Value added
243	−4.04	−6.35	−4.53	−14.00
313	−0.09	−4.04	−1.21	−0.98
333	−6.08	−12.30	−7.55	−0.57
241	−1.68	−3.35	−2.24	−0.12
151	3.12	0.40	2.44	0.32
152	3.34	−1.14	1.81	0.94
251	1.29	−0.53	0.87	2.29
160	−0.88	−0.91	−0.88	2.49
154	0.47	−1.22	0.13	2.66
353	4.30	−0.36	2.37	3.59

Note Industries entered in the ascending order in terms of value added growth rate
Source Annual Survey of Industries, (ASI)

with sluggish growth in terms of value added has been either negative or sluggishly positive (Table 3). Overall, industries not performing well in terms of value added did not perform well in terms of employment either.

There is a strong positive correlation between the average value added growth and total employment growth measured across all the three-digit manufacturing groups (0.77), implying growth is essential for employment generation. However, not necessarily rapid value added growth has resulted in faster employment growth. In spite of the fact that many industries grew rapidly in value added terms total employment increased only at around 2.6 % per annum at the aggregate level over the period 1998–1999 through 2007–2008. However, during the sub-period 2003–2004 through 2008–2009 as Goldar (2011a, b) pointed out, employment and value added both grew sizably.

The list of industries with a rapid employment growth (i.e. of at least 4 % per annum) and a rapid value added growth (i.e. of at least 7 % per annum) over 1998–1999 to 2007–2008 includes 155 (manufacture of beverages), 369 (manufacturing n.e.c.), 332 (manufacture of optical instruments etc.), 372 (recycling of non-metal waste and scrap), 312 (manufacture of electricity distribution and control apparatus), 191 (tanning and dressing of leather, handbags), 273 (casting of metals), 252 (manufacture of plastic products), 361 (manufacture of furniture), 371 (recycling of metal waste and scrap), others, 289 (manufacture of other fabricated metal products), 181 (manufacture of wearing apparel), 343 (manufacture of parts for motor vehicles and their engines), 173 (manufacture of knitted and crocheted fabrics), 172 (manufacture of other textiles), 281 (manufacture of structural metal products, tanks etc.), 319 (manufacture of other electrical equipment), 300 (manufacture of office, accounting and computing machinery), 232 (Manufacture of refined petroleum products), 182 (Dressing and dyeing of fur etc.), 269 (manufacture of non-metallic mineral products), 192 (manufacture of footwear etc.): see the box below.

Industries with Rapid Value Added Growth and Employment Growth
More than 7 % growth in value added (in ascending order)
223, 210, 261, 155, 242, 351,361, 181, 369, 202, 342, 322, Total, 272, 323,
271,291, 172, 269,289, 292, 315, 182,343, 319, 371, 332, 314, 359,331, 173,
281, 311,, 312, 231, Others, 341, 300, 232, 372
More than 4 % employment growth rate (in ascending order)
273, 191, 361, 155, 252, 232, 182, 332, 269, 312, 300, 281, 192, 289, 319,
Others, 343, 369, 371, 181, 173, 172, 372
Source: See Table 1.

Decomposing total employment in terms of workers and employees (other than
workers), several industries are seen to have experienced a negative growth rate for
the latter though value added growth in these industries has been more than 5 % per
annum (Table A.2 in the appendix). Rise in capital intensity and skill intensity in
the organized manufacturing in India is a widely noted phenomenon. In the
backdrop of this, the sluggish growth of 1.3 % per annum in the number of
employees (other than workers) constituting administrative, managerial and other
skilled workers comes as a major surprise. Possibly the firms had to cut down the
number of skilled workers as they are increasingly becoming expensive with the
rise in the skill premium, almost catching up with the international standards.
A weak correlation, though positive, observed between the growth in employees
other than workers and the growth in salary per employee is possibly indicative of
this fact. There is in fact a great deal of the literature to suggest that the skill
premium in the developing countries is very high which has resulted in increasing
inequality in terms of remunerations for different categories of labour (Mitra 2013).

Industries which recorded a negative growth in terms of total employment also
witnessed a negative growth in terms of employees other than workers with the
exception of 201 (saw milling etc.): see Table A.3 in the appendix. The growth rate
in terms of workers has also been negative in most of these industries.[15] The
correlation between the growth rates in workers and employees other than workers
turns out to be high at 0.91.

Among the industries which grew sluggishly (0–2 %) in terms of total
employment several witnessed negative growth rate in terms of employees other
than workers but positive in terms of workers. And among them many again
reported a rapid value added growth rate (Table A.4 in the appendix).

Industries which experienced an employment growth of 2–4 (or marginally
higher than that) percent per annum, workers growth is seen to be mostly greater

---

[15]Except 291 (manufacture of general purpose machinery) and 359 (manufacture of transport
equipment). Interestingly the value added growth rate has been positive except in the case of 333
(manufacture of watches and clocks), 243 (manufacture of man-made fibres), 241 (manufacture of
basic chemicals) and 313 (manufacture of insulated wire and cables).

Table 4 Value added and total employment growth (% per annum)

Employment growth rate	Value added growth rate			
	Negative/Sluggish (less than 4 % p.a.)	Moderate (4 and above but less than 7 % p.a.)	High (7 % p.a. and above)	Total
Negative/sluggish (less than 2 % p.a.)	152, 154, 160, 241, 243, 251, 313, 333	153, 171, 201, 221, 293, 352	223, 261, 272, 291, 311, 314, 322, 323, 341, 351, 359	25
Moderate (2 and above but less than 4 % p.a.)	151, 353	222, 321	202, 210, 231, 242, 271, 292, 315, 331,342, aggregate	14
High (4 % p.a. and above)	0	191, 192, 252, 273	155, 172, 173, 181, 182, 232, 269, 281, 289, 300, 312, 319, 332 343, 361, 369, 371, 372, others	23
Total	10	12	39	61

than the growth in employees other than workers. Also, the value added growth has been highly impressive in a number of industries in this category: Table A.5 in the appendix.

Among the rapidly growing industries, i.e. more than 4 % in terms of total employment, workers growth has been substantially higher than the growth in employees other than workers (Table A.6 in the appendix).[16] Also, the value added growth rate has been fast in most of the industry groups in this category.

Four clusters are discernible from Tables 4 and 5: (a) low employment and low value added growth, (b) low employment but high value added growth, (c) moderate employment and high value added growth and (d) high employment and high value added growth (also see the Figs. 1 and 2).

On the whole, we observe that for the entire period under consideration (1998–1999 through 2007–2008) value added growth has been fast in a number of industries. However, employment growth not necessarily has been impressive in these industries though rapidly growing industries in terms of employment witnessed faster value added growth as well. In fact, in some of the industries with a marginal or sluggish employment growth, value added still has grown sizably notwithstanding a strong positive correlation between the value added and employment (average) growth rates across industries. Particularly, the growth scenario of employees other than workers represents a gloomy picture since many industries showed a negative growth rate. This comes as a bit of surprise, particularly keeping in view the popular belief about a favourable job market for the ones

---

[16]The exceptions are 182 (dressing and dyeing of fur etc.), 332 (manufacture of optical instruments etc.), others, 369 (manufacturing n.e.c.), 371 (recycling of metal waste and scrap) and 181 (manufacture of wearing apparel), 172 (manufacture of other textiles) and 372 (recycling of non-metal waste and scrap).

Table 5 Value added and workers growth (% per annum)

Growth rate in workers	Value added growth rate			
	Negative/sluggish (less than 4 % p.a.)	Moderate (4 and above but less than 7 % p.a.)	High (7 % p.a. and above)	Total
Negative/sluggish (less than 2 % p.a.)	154, 160, 241, 243, 251, 313, 333	153, 171, 201, 221, 293, 352	223, 261, 272, 291, 314, 322, 323, 359	21
Moderate (2 and above but less than 4 % p.a.)	151, 152,	222, 321,	202, 210, 231, 242, 271, 292, 311, 315, 341, 351, aggregate	14
High (4 % p.a. and above)	353	191, 192, 252, 273,	155, 172, 173, 181, 182, 232, 269, 281, 289, 300, 312, 319, 331, 332, 342, 343, 361, 369, 371, 372, others	26
Total	10	12	39	61

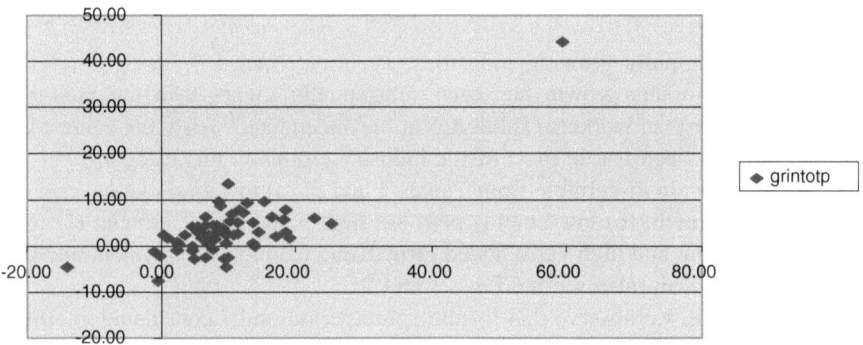

Fig. 1 Scatter of growth rate in value added and total employment (grintotp)

who are highly skilled. Usually greater concern has been expressed for the unskilled workers as they are characterized by poor employability. Two reasons may be considered to explain this: (a) because of a high level of salary for the employees other than workers their absorption rate has been sluggish, (b) as mentioned in the previous section the recent phase of industrialization is partly because of the rapid spread of industries in the states which were less industrialized earlier and hence, this spur has been accompanied by a rise in the demand for shop floor workers. Nevertheless, there are a sizeable number of industries which experienced rapid growth in terms of value added and total employment both.

An important question which needs to be addressed is whether the so-called labour-intensive industries have been generating employment significantly. Fixed

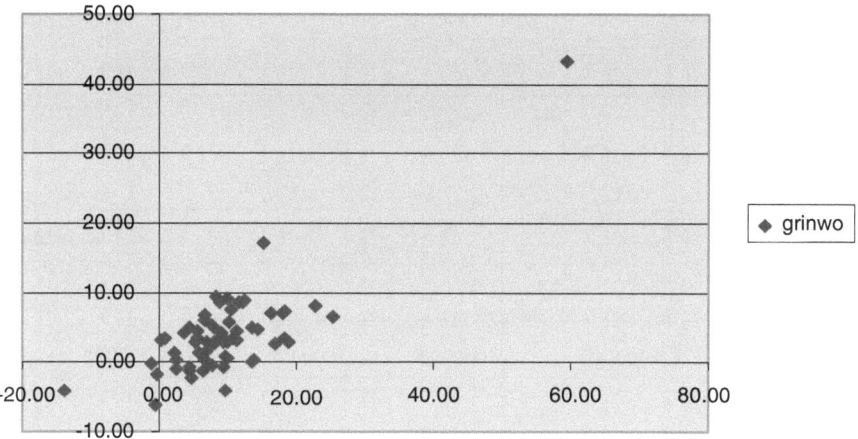

Fig. 2 Scatter of growth rate in value added and workers (grinwo)

Table 6 Employment and capital growth

Dependent variable	Independent variable	Coefficient (t-ratio)	R2
Employment growth	Rate of growth in capital-labour ratio	0.32 (2.46)*	0.08
Employment growth	Capital-labour ratio	0.0000006 (−0.77)	0.009

Note The average rate of growth in employment over 1998–1999 through 2007–2008 has been regressed on the average rate of growth in capital-labour (total employment) ratio and also the average capital-labour (total employment) ratio. Number of observations is 62
*Represents significance at 1 per cent level

capital and employment growth show a strong positive correlation of 0.82 across various industry groups. There is a positive relationship between the rate of growth in capital-labour (i.e. total person engaged) ratio and employment growth, implying both the factors of production can increase simultaneously though capital may be increasing at a faster pace than labour (Table 6). We also note that higher is the level of capital–labour ratio, lower is the employment growth rate though the relationship is not statistically significant, implying while some of the labour intensive industries may be experiencing rapid employment growth some others tend to grow sluggishly (see Fig. 3).

However, there are some labour-intensive industries which showed a rapid employment growth rate. Among the 32 industries placed at the bottom in terms of capital–labour ratio (i.e. total persons engaged) nearly half reported a faster employment growth rate of more than 4 % per annum as listed below.[17]

---

[17]Similarly among the top twenty capital-intensive industries 269, 155, 300, others, and 232 registered a rapid employment growth of more than 4 % per annum.

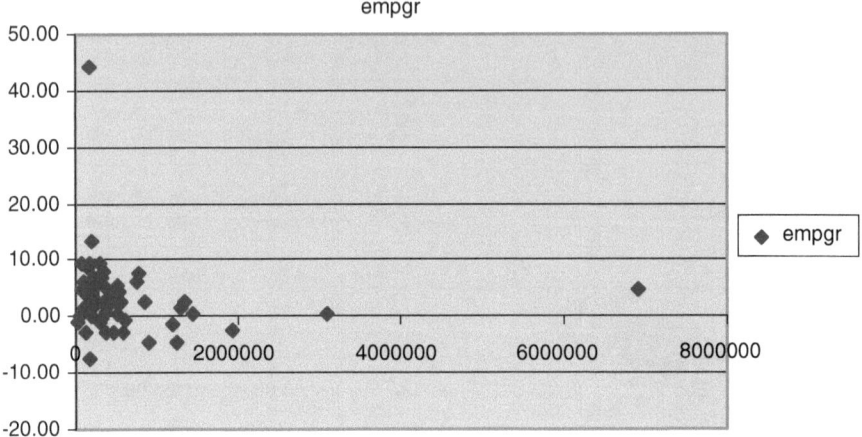

Fig. 3 Capital–labour (total employment) ratio and employment growth rate (emp gr)

181, 182, 192, 191, 369, 281, 372, 371, 172, 361, 273, 289, 312, 173, 319, 343 (in ascending order of capital-labour ratio)

3.2.1 Econometric Estimation of Employment Elasticity

Employment elasticity is defined as the proportionate change in employment due to the proportionate change in value added. This, however, needs to be estimated econometrically after controlling the remuneration to the human capital. On the whole, we are able to observe the partial elasticity of employment with respect to growth and that with respect to remuneration rate.

The employment elasticity function has been estimated separately for total persons engaged and workers (Table A.7 in the appendix). The logarithm transformation of total persons engaged (and separately for workers) has been regressed on logarithm transformation of total value added and emoluments per person (and wages per worker in the specification for workers). Both the growth elasticity and wage/remuneration elasticity have been estimated to assess the growth and wage sensitivity of employment. For the aggregate manufacturing sector, the elasticity of total employment with respect to value added is 0.43 and the elasticity of workers with respect to value added is 0.35. On the other hand, the impact of remuneration per employee (or wages per worker) on total employment (or workers) is not statistically significant. Also, at the disaggregated level for a number of industries the elasticity of total employment with respect to growth does not turn out to be statistically significant.[18] Industries which recorded an employment elasticity of up to 0.4 are many[19] though only a handful of them had an employment elasticity of more than 0.55 (given in ascending order).

342 (manufacture of bodies for motor vehicles etc.), 289 (manufacture of other fabricated metal products), 371 (recycling of metal waste and scrap), 223 (reproduction of recorded media), 273 (casting of metals), 192 (manufacture of footwear etc.), 182 (dressing and dyeing of fur etc.), 359 (manufacture of transport equipment), 269 (manufacture of non-metallic mineral products), 372 (recycling of non-metal waste and scrap), 331 (manufacture of medical appliances and instruments), 173 (manufacture of knitted and crocheted fabrics) and 172 (manufacture of other textiles). Though one may argue that these industries may be targeted for providing a boost to employment growth, it is unlikely that they will succeed in raising the overall employment growth of the manufacturing sector in a significant manner as specified in the NMP.

3.2.2 Decomposition of Value Added: Productivity and Employment

Decomposition of value added growth in terms of employment growth and productivity growth is carried out to delineate the contribution of labour to value added growth vis-à-vis the capital-intensive technology (Table A.8 in the appendix). Only a handful of industries are seen to have experienced a rapid productivity growth of

---

[18]152 (manufacture of dairy product), 154 (manufacture of other food products), 155 (manufacture of beverages), 160 (manufacture of tobacco products), 201 (Saw milling and planing of wood etc.), 202 (manufacture of products of wood, cork etc.), 243 (manufacture of man-made fibres), 252 (manufacture of plastic products), 261 (manufacture of glass and glass products), 292 (manufacture of special purpose machinery), 293 (manufacture of domestic appliances), 311 (manufacture of electric motors, generators and transformers), 313 (manufacture of insulated wire and cables), 314 (manufacture of accumulators, primary cells and batteries), 315 (manufacture of electric lamps etc.), 319 (manufacture of other electrical equipment), 321 (manufacture of electronic valves and tubes etc.), 322 (manufacture of television and radio transmitters), 333 (manufacture of watches and clocks), 343 (manufacture of parts for motor vehicles and their engines), 351 (building and repair of ships and boats), 352 (manufacture of railway and tramway locomotives), 361 (manufacture of furniture) and other.

[19]231 (manufacture of coke oven products), 332 (manufacture of optical instruments etc.), 353 (manufacture of aircraft and spacecraft), 232 (manufacture of refined petroleum products), 153 (manufacture of grain mill products etc.), 221 (publishing), 191 (tanning and dressing of leather, handbags), 242 (manufacture of other chemical products), 323 (manufacture of television and radio receivers, recording apparatus), 151 (production of meat, fish, fruit, vegetables, etc.), 300 (manufacture of office, accounting and computing machinery), 222 (printing and service related to printing), 312 (manufacture of electricity distribution and control apparatus), 272 (manufacture of basic precious and non-ferrous metals), 291 (manufacture of general purpose machinery), 369 (manufacturing n.e.c.), 251 (manufacture of rubber products), 171 (spinning, weaving and finishing of textiles) and 181 (manufacture of wearing apparel).

Table 7 Industries with rapid productivity, worker and wage growth

VA per worker (5 % and more)	Worker (4 % and more)	Wage (2 % and more)
171, 182, 201, 202, 221, 223, 231, 232, 242, 261, 271, 272, 281, 291, 292, 293, 300, 311, 312, 314, 315, 322, 323, 331, 332, 333, 341, 351, 352, 359, 372, others, total	155, 172, 173, 181, 182, 191, 192, 232, 252, 269, 273, 281, 289, 300, 312, 319, 331, 332, 342, 343, 353, 361, 369, 371, 372, others	172, 181, 182, 201, 251, 272, 323, 332, 353, 359, 369, 371, 372, others

at least 5 % per annum and employment growth of at least 4 % per annum simultaneously.[20] Though there is no huge trade-off (in terms of a negative correlation) between productivity growth and employment growth across industries, the positive correlation across industries is highly negligible, i.e. 0.08. In other words, capital-intensive technology has been adopted in several industries, leading to a rapid value added growth and labour productivity growth. This is also reflected in the growth in capital–labour ratio across various groups that witnessed a growth rate of more than 4 % per annum.[21]

The correlation between the average growth in workers' productivity and wages is again negligible (0.07). Table 7 lists the industry codes which registered a rapid growth rates in workers' productivity (value added per worker, at least 5 % per annum), worker (at least 4 % per annum) and wages (at least 2 % per annum). The following observations are pertinent:

1. Only 182 (Dressing and dyeing of fur etc.), 332 (Manufacture of optical instruments etc.), 372 (Recycling of non-metal waste and scrap) and 'others' show a rapid growth in terms of all the three characteristics simultaneously.

---

[20]173 (manufacture of knitted and crocheted fabrics), 182 (dressing and dyeing of fur etc.), 232 (manufacture of refined petroleum products), 281(manufacture of structural metal products, tanks etc.), 300 (manufacture of office, accounting and computing machinery), 312 (manufacture of electricity distribution and control apparatus), 319 (manufacture of other electrical equipment), 332 (manufacture of optical instruments etc.), 372 (recycling of non-metal waste and scrap) and others.

[21]in ascending order of magnitude: 160 (manufacture of tobacco products), 261 (manufacture of glass and glass products), 292 (manufacture of special purpose machinery), 273 (casting of metals), 153 (manufacture of grain mill products etc.), 243 (manufacture of man-made fibres), 352 (manufacture of railway and tramway locomotives), 231 (manufacture of coke oven products), 351 (building and repair of ships & boats), 154 (manufacture of other food products), 311 (manufacture of electric motors, generators and transformers), 281 (manufacture of structural metal products, tanks etc.), 221 (publishing), 353 (manufacture of aircraft and spacecraft), 173 (manufacture of knitted and crocheted fabrics), 359 (manufacture of transport equipment), 314 (Manufacture of accumulators, primary cells and batteries), 201 (Saw milling and planing of wood etc.), 323 (Manufacture of television and radio receivers, recording apparatus) (Manufacture of television and radio receivers, recording), 291(Manufacture of general purpose machinery), 182 (Dressing and dyeing of fur etc.), 300 (Manufacture of office, accounting and computing machinery), 322 (Manufacture of television and radio transmitters), Others and 372 (Recycling of non-metal waste and scrap), and 371 (Recycling of metal waste and scrap).

2. Industries in which rapid growth took place simultaneously in terms of labour productivity (value added per worker, at least 5 % per annum) and wage per worker (at least 2 % per annum) are 201 (saw milling and planing of wood etc.), 272 (manufacture of basic precious and non-ferrous metals), 323 (manufacture of television and radio receivers, recording apparatus) and 359 (manufacture of transport equipment).
3. Industries in which rapid growth took place simultaneously in terms of worker (at least 4 % per annum) and wage per worker (at least 2 % per annum) include 172 (Manufacture of other textiles), 181 (Manufacture of wearing apparel), 353 (Manufacture of aircraft and spacecraft), 369 (Manufacturing n.e.c.) and 371 (Recycling of metal waste and scrap). However, the correlation between the growth rate in workers and that in the wage rate per worker across is not negligible (0.54).
4. Industries with rapid growth in value added per worker (at least 5 % per annum) and worker (at least 4 % per annum) include: 232 (Manufacture of refined petroleum products), 331 (Manufacture of medical appliances and instruments), 332 (Manufacture of optical instruments etc.), 372 (Recycling of non-metal waste and scrap) and others. The correlation between the growth rates of these two variables across industries turns out to be only 0.12 indicating capital accumulation led productivity growth.

3.3 Labour Market Flexibility and Wages and Salaries from ASI Data

3.3.1 Background

Sluggish employment growth or the absence of significant positive effects of economic growth on living standards through productive employment generation is viewed as an outcome of strict labour market regulations. Some of the studies indicate that flexibility in the labour market and reforms in other sectors of the economy are expected to raise employment and also the real wages in the long-run, if not in the short-run (Fallon and Lucas 1991). Hence, there has emerged a strong case for speedy reforms in the labour market together with reforms in other segments of the economy. But it is a well-known fact that in the process of economic reforms there are certain social costs particularly for the existing labour force, and any such reforms without addressing these costs would only have adverse impact on the labour market. Further, the ILO-SAAT study (ILO-SAAT 1996) emphasizes that although labour market reforms could reduce the social costs of structural adjustments, it cannot minimize the social costs of stabilization. Though deregulation might be desirable in the context of increasing integration of a particular economy with the global economy, any reform in the labour market, the study argues, needs to be a gradualist one and it has to be accompanied by appropriate social safety nets and some institutional innovations that help the job losers in the process of labour reallocation. Hasty reforms would only reduce the social welfare. Moreover, as the organized labour market in several developing countries like India account for only a small fraction of the work force, "it is legitimate to wonder if labour market reform can be of much consequence in such context" (ILO-SAAT 1996).

Goldar and Aggarwal (2005) noted an accelerated fall in the income share of labour in the manufacturing sector in India during the post-reform period (also see Datta 2003; D'souza 2008). Hasan et al. (2003) bring out that labour demand elasticities increase with reductions in protection, and further, Indian states with more flexible labour markets[22] witnessed larger increases in labour demand elasticities in response to reductions in protection, highlighting the importance of institutional factors. In fact in his study of 48 developing countries Hasan (2003), using panel data, argue that trade liberalization is more likely to have a beneficial impact on employment when labour markets are flexible and vice versa. More regulated and rigid labour markets, he noted, are associated with higher real wages, which, however, come at the expense of employment. The bargaining power provides strong support only to those who are employed in the formal sector of the economy, constituting an insignificant component of the work force. Thus according to him bargaining power does not support expansion of employment opportunities. Evidence on limited benefits from trade liberalization for the typical worker largely refers to the Latin American experience. And some argue that trade does have the potential to benefit workers at large though the nature of labour market regulations actually play an important role in giving a tangible shape to these benefits (Edwards and Edwards 1994). Hence, one important view, as mentioned above, is that with the presence of regulated labour market, the overall impact of economic policy may not necessarily have positive impact on employment generation. Besley and Burgess (2004) show that Indian states which amended the Industrial Disputes Act in a pro-worker direction experienced lowered output, employment, investment and productivity in organized manufacturing. Bhattacharjea (2006) on the other hand criticizes the widely used index of state-level labour regulation devised by Besley and Burgess (2004), and the econometric methodology they use to establish that excessively pro-worker regulation led to poor performance in Indian manufacturing. On the whole, while there may be a case for removing labour market rigidities by discouraging the political patronization of the unions and relaxing the strict labour laws that prohibit employment growth, attention also needs to be given to the labour welfare issues.

Since much of the focus on labour market deregulation lies on introducing wage flexibility or in other words removal of downward stickiness of wages so as to expand employment opportunities in the high productivity sector, we may like to know how responsive employment is in relation to wages in the manufacturing sector.

---

[22]In their study a state is said to have flexible labour market if the state had undertaken anti-employee amendments in the Industrial Dispute Act.

Mitra (2013) noted that the elasticity of employment with respect to wages across many countries is quite low even when it takes the right sign (negative). This piece of evidence may be taken to suggest that deregulation in terms of wage flexibility does not have a strong standing across countries. In other words, raising employment substantially through wage reduction does not seem to be promising in a large majority of the developing countries.

Those who support labour market deregulation believe that the cost of labour is too high, thus urging that in the context of globalization firms need to become more competitive by cutting cost. In this sense, high labour cost is thought to be one of the major sources of inefficiency. Labour rules and strong unions are believed to push the wage rate artificially much above the market clearing wage rate which in turn is said to reduce employment (Fallon and Lucas 1991). Hence, labour market deregulation is expected to reverse the attitude of the employers against expanding employment since it empowers them to hire and fire labour as per requirement and offer wages which allow product prices to remain competitive.

Ahsan and Pages (2009) using manufacturing sector data for India, brings out the economic effects of legal amendments on two types of labour laws: employment protection and labour dispute resolution legislation. Laws that increase employment protection or the cost of labour disputes substantially reduce registered or organized sector employment and output or the share of value added that goes to labour. Labour-intensive industries, such as textiles, are the hardest hit by amendments that increase employment protection. On the other hand, the capital-intensive industries are affected by laws that increase the cost of labour dispute resolution. These adverse effects particularly relating to employment, are not alleviated by the widespread and increasing use of contract labour. Besley and Burgess also (2004) showed that Indian states which amended the Industrial Disputes Act in a pro-worker direction experienced lowered output, employment, investment and productivity in the organized manufacturing.

Mitra (2013) examined if there is any connection between labour absorption in manufacturing sector and labour market regulation across countries.[23] World Development Indicators report the percentage of managers indicating labour regulations (LABREG) as a major business constraint and the percentage of managers indicating labour skill as a major business constraint (LABSKILL) for various countries. Higher is the percentage, higher is the probability that labour market regulations and skill factor affect employment adversely. We have tried to relate these skill and regulation-specific responses to the ratio of labour to real value added (LTORVA) estimated from UNIDO data for the aggregate manufacturing sector. The coefficients in the estimated equations, however, turn out to be highly insignificant. Even when we control for real wage rate in the manufacturing sector

---

[23]Berg and Cazes (2007) point out the serious conceptual and methodological problems associated with the World Bank's Employing Workers Index of the Doing Business indicators and risks of formulating policies on the basis of these indicators.

(RWAGE), GDP per capita (GDPPC) at the national level and the share of manufactures in total imports (MFGIM) taken as a proxy for imported technology, neither the labour skill variable nor labour market regulation turns out to be significant. Alternate estimate of labour absorption (or dependent variable) have been tried in the equation, i.e. the rate of growth of employment (ROGEMFG) in the manufacturing sector from UNIDO data. Interestingly the skill factor is seen to affect employment growth in the manufacturing sector negatively. In other words, higher is the percentage of managers who feel skill has been affecting business adversely, lower is the rate of growth of employment in the manufacturing sector. This implies that poor skill base of the work force in the developing countries reduces the pace of labour absorption as labour demand is possibly rising only for the high skilled variety. However, the effect of the labour market regulation is not statistically significant on the alternate form of the dependant variable. Though the sample is quite small, at least this much is evident that labour market regulations do not retard labour absorption. ILO's data on the percentage of workers registered with the unions can be tried as a proxy for labour market condition though there are serious problems.[24]

Below, we present certain broad patterns emerging from the ASI data for India.

3.3.2 Wages and Salaries: Evidence

Wage share, salary share and total emolument share have been worked out for each of the industries on year to year basis (Table A.9 in the appendix). The average emolument share at the aggregate level is around 20 %, with 50 % each to workers and employees other than workers. In a large majority of the industries it is below 30 %.[25] Since wages do not constitute a large component of the total cost a more careful analysis has to be pursued to identify what restricts industrial expansion and employment creation rather than simply blaming the labour laws.

The wage–salary inequality is very much distinct from Table A.9 in the appendix. The average ratio of wage per worker to salary per employee (other than worker) is only around 0.3 at the aggregate level: at the disaggregated level in

---

[24]The ratio of trade union members as a percentage of total paid employees has been calculated by ILO Bureau of Statistics.

[25]Only corresponding to the following industries it has been more than 30 %: 152 (Manufacture of dairy product), 154 (Manufacture of other food products), 191 (Tanning and dressing of leather, handbags), 192 (Manufacture of footwear etc.), 201 (Saw milling and planing of wood etc.), 221 (Publishing), 222 (Printing and service related to printing), 243 (Manufacture of man-made fibres), 273 (Casting of metals), 333 (Manufacture of watches and clocks), 342 (Manufacture of bodies for motor vehicles etc.), 343 (Manufacture of parts for motor vehicles and their engines), 351 (Building and repair of ships and boats), 352 (Manufacture of railway and tramway locomotives), 353 (Manufacture of aircraft and spacecraft), 361 (Manufacture of furniture), 371 (Recycling of metal waste and scrap) and 372 (Recycling of non-metal waste and scrap). Among them most are, however, below 40 % with the exception of a few.

almost two-third of the industries the ratio turns out to be more than 0.4.[26] On the other hand, the worker to employees (other than workers) ratio is almost 3.4 times at the aggregate level. All this would tend to indicate that wages are not necessarily placed at a high level which in turn could result in sluggish industrial employment for relatively less-skilled labour, as popularly believed. Hence, the arguments for labour market deregulation and reduction in wages may not turn out to be effective in boosting employment sizably. In fact, the wage elasticity of workers (i.e. elasticity of workers with respect to wage rate) turns out to negative and significant only in a group of 16 industries as noted from Table A.7 in the appendix, which reports the partial elasticity of workers with respect to value added and wages.

Wage share can be interpreted as the wage rate to productivity ratio. Rise in wage share would, therefore, mean a greater increase in the wages compared to labour productivity growth. An increase in wage share over time can be interpreted as a rise in inefficiency. However, while interpreting this as a rise in inefficiency we need to be careful if wages to begin with were placed at a too low level.

The following are the industries in which we witnessed a more than 1 % increase in real wage rate which has also been more than the growth rate in value added per worker: 151, 251, 172, 369, 181, 353 and 371.[27] These industries are, however, too few in number, implying that the argument of wages being pushed to a high level that exceeds worker productivity and introduces inefficiency into the system cannot

---

[26]152 (Manufacture of dairy product), 272 (Manufacture of basic precious and non-ferrous metals), others, 351 (Building and repair of ships and boats), 372 (Recycling of non-metal waste and scrap), 371 (Recycling of metal waste and scrap, 231 (Manufacture of coke oven products), 221 (Publishing), 361 (Manufacture of furniture), 153 (Manufacture of grain mill products etc.), 291 (Manufacture of general purpose machinery), 353 (Manufacture of aircraft and spacecraft), 172 (Manufacture of other textiles) and 201 (Saw milling and planing of wood etc.) (in the ascending order). On the other hand, the number of workers is almost three times that of the salary holders in very many industries: 155 (Manufacture of beverages), 271 (Manufacture of Basic Iron and Steel), 222 (Printing and service related to printing), 343(Manufacture of parts for motor vehicles and their engines), 201 (Saw milling and planing of wood etc.), 151 (Production of meat, fish, fruit, vegetables, etc.), 289 (Manufacture of other fabricated metal products), 251 ((Manufacture of rubber products), 202 (Manufacture of products of wood, cork etc.), 153 (Manufacture of grain mill products etc.), 333 (Manufacture of watches and clocks), 359 (Manufacture of transport equipment), 273 (Casting of metals), 315 (Manufacture of electric lamps etc.), Total, 351 (Building and repair of ships and boats), 210 (Manufacture of paper and paper product), 243 (Manufacture of man-made fibres), 231 (Manufacture of coke oven products), 369 (Manufacturing n.e.c.), 172 (Manufacture of other textiles), 154 (Manufacture of other food products), 173 (Manufacture of knitted and crocheted fabrics), 261 (Manufacture of glass and glass products), 372 (Recycling of non-metal waste and scrap), 269 (Manufacture of non-metallic mineral products), 191 (Tanning and dressing of leather, handbags), 342 (Manufacture of bodies for motor vehicles etc.), 371 (Recycling of metal waste and scrap), 192 (Manufacture of footwear etc.), 182 (Dressing and dyeing of fur etc.), 171 (Spinning, weaving and finishing of textiles), 181 (Manufacture of wearing apparel) and 160 (Manufacture of tobacco products) (in ascending order).

[27]In ascending order of wage growth rate.

be held strongly against the Indian manufacturing sector. Even if we take simply the ratio of wages per worker to labour productivity not a single case is seen where the numerator has been more than the denominator. Thus, the popular views on unionization and inefficiency are apparently exaggerated.

3.3.3   Regular Workers to Contractual Workers

The process of contractualization is on the rise even at the aggregate level as brought out by the NSS data. The percentage of work force engaged as casual workers has shot up in 2009–2010 compared to 2004–2005. Even within the ASI sector, the proportion of contractual workers to total workers (directly employed plus those appointed through the contractors) has increased steadily over time. Second, in many industries the proportion has been extremely high. The third observable pattern relates to the ratio of female to male workers in the category of directly recruited workers, which is significant in a group of industries. Since female wages are usually believed to be lower than that of the males, feminization of the work force may be treated as an indicator of labour market flexibility that the employers have chosen as a means to reduce labour cost. The following industries show a more than 200 female workers per 1000 male workers in 2007–2008 (see the box below):

191 (Tanning and dressing of leather, handbags), 322 (Manufacture of television and radio transmitters), 300 (Manufacture of office, accounting and computing machinery), 210 (Manufacture of paper and paper product), 151 (Production of meat, fish, fruit, vegetables etc.), 332 (Manufacture of optical instruments etc.), 182 (Dressing and dyeing of fur etc.), 315 (Manufacture of electric lamps etc.), 242 (Manufacture of other chemical products), 321 (Manufacture of electronic valves and tubes etc.), 201 (Saw milling and planing of wood etc.), 223 (Reproduction of recorded media), 173 (Manufacture of knitted and crocheted fabrics), 192 (Manufacture of footwear etc.), 154 (Manufacture of other food products), 333 (Manufacture of watches and clocks), 160 (Manufacture of tobacco products) and 181 (Manufacture of wearing apparel).

At the aggregate level for the year 2007–2008 the ratio of workers appointed through contractors to those directly employed by the firms turns out to be 0.44 which increased steadily from 0.36 in 2004–2005, implying rapid contractualization within a short time span. In the following industries, the ratio has been more than 0.5 (see the box below).

289 (Manufacture of other fabricated metal products), 293 (Manufacture of domestic appliances), 315 (Manufacture of electric lamps etc.), 312 (Manufacture of electricity distribution and control apparatus), 313 (Manufacture of insulated wire and cables), 152 (Manufacture of dairy product), 300 (Manufacture of office, accounting and computing machinery), 273 (Casting of metals), 271 (Manufacture of Basic Iron and Steel), 321 (Manufacture of electronic valves and tubes etc.), 343 (Manufacture of parts for motor vehicles and their engines), 241 (Manufacture of basic chemicals), 153 (Manufacture of grain mill products etc.), 151 (Production of meat, fish, fruit, vegetables, etc.), 182 (Dressing and dyeing of fur etc.), 352 (Manufacture of railway and tramway locomotives), 322 (Manufacture of television and radio transmitters), 155 (Manufacture of beverages), 281 (Manufacture of structural metal products, tanks etc.), 269 (Manufacture of non-metallic mineral products), 342 (Manufacture of bodies for motor vehicles etc.), 351 (Building and repair of ships and boats), 232 (Manufacture of refined petroleum products), 160 (Manufacture of tobacco products) and 372 (Recycling of non-metal waste and scrap).

In fact, some of the industries[28] show a ratio of more than unity, implying that the percentage of contractual workers in the total was more than 50 %, and then going up to almost 73 %. It is important to know if the industries which recorded a relatively higher proportion of contractual workers in general, also experienced a faster growth in total employment or total workers. In a couple of industries, we observe that the growth in workers has been more than 4 % per annum over the period 1998–1999 through 2007–2008 and at the same time the contractual to directly employed workers ratio has been more than 0.5 (i.e. the percentage of contractual workers being more than one third of the total workers) in 2007–08 (see the Box below): Since the number of such industries is not too large, a strong case cannot be made for contractualization as a key to higher employment growth rate. Also, the wage elasticity of employment (i.e. the responsiveness or the elasticity of worker with respect to wage per worker) is found to be high only in a few cases (Table A.7), again indicating that wage flexibility in the downward direction need not raise the employment substantially. As D'souza (2008) points out, even without deregulating the labour market the employers already have the freedom to choose the mix of regular and contractual workers, which enables them to have control over labour and reduce the wage cost in a significant manner.

---

[28]322 (Manufacture of television and radio transmitters), 155 (Manufacture of beverages), 281 (Manufacture of structural metal products, tanks etc.), 269 (Manufacture of non-metallic mineral products), 342 (Manufacture of bodies for motor vehicles etc.), 351 (Building and repair of ships and boats), 232 (Manufacture of refined petroleum products), 160 (Manufacture of tobacco products) and 372 (Recycling of non-metal waste and scrap).

155 (Manufacture of beverages), 182 (Dressing and dyeing of fur etc.), 232 (Manufacture of refined petroleum products), 269 (Manufacture of non-metallic mineral products), 273 (Casting of metals), 289 (Manufacture of other fabricated metal products), 300 (Manufacture of office, accounting and computing machinery), 312 (Manufacture of electricity distribution and control apparatus), 342 (Manufacture of bodies for motor vehicles etc.), 343 (Manufacture of parts for motor vehicles and their engines) and 372 (Recycling of non-metal waste and scrap).

4 Performance of Unorganized Manufacturing

In this chapter, we turn to the unorganized manufacturing sector and analyze its performance during the period 1989–1990 through 2005–2006 and 2005–2006 through 2009–2010. No time series information is available on the unorganized manufacturing enterprises except the point estimates available from the special surveys conducted by the NSS Organization. The unregistered manufacturing sector, which is taken as an operational synonym for the unorganized manufac-turing sector, includes units which employ less than 10 workers and units which employ 10–19 workers but not using electricity or power. The unregistered man-ufacturing as per the National Accounts Statistics of the Central Statistical Organisation accounts for around 34 % of the total manufacturing value added.

The units within the unorganized manufacturing sector have been divided into three types: own account manufacturing enterprises (OAMEs) are those which use only household or family labour, non-directory manufacturing enterprises (NDMEs) employ 1–5 workers of which at least one is hired and the directory manufacturing enterprises (DMEs) in the unregistered manufacturing include units with 6–9 workers irrespective of using power, and units with 10–19 workers without using power. However, the definition of workers in the surveys on unor-ganized manufacturing enterprises by NSSO is very broad. No distinction is made between fulltime and part-time workers, and more importantly no time dimension is used in defining a worker. In other words, anyone attached to the unit in whatever way possible, is defined as a worker. The interpretation of employment-related concepts in this sector, therefore, has to be made very carefully. Further, we may note that the recent survey, 2010–2011 has not provided the data for NDMEs and DMEs separately—all being clubbed under establishments.

4.1 *Growth in Employment and Output*

Growth in employment and output are undoubtedly two important indicators of performance of the industrial sector. It would be useful therefore to compare growth rates in employment and output attained in the reform period.

Table 8 presents a comparison of growth rates in terms of employment, output and number of enterprises for the OAME, NDME and DME—the three segments of the unorganized manufacturing component. It is seen that growth in the reform period has been relatively faster in NDME segment compared to the other two segments, particularly in terms of employment and number of enterprises. However, for the aggregate unorganized manufacturing the growth in employment has been extremely sluggish.

Analysis of output growth in unorganized manufacturing by major industry groups shown in Table 8 reveals that textiles and leather, non-metallic mineral products, basic metals, metal products, and machinery and transport equipment achieved relatively faster growth in real value added compared to the other sectors during the post-reform period. However, employment growth turned out to be as high as 2 % per annum only in textiles, chemical, metal products and transport equipment (Table 8).

Almost uniformly, growth in urban areas is found to be faster. The only exception is growth in real value added in NDME and DME—the growth rate in rural areas exceeded that in the urban areas though the total value added growth in the unorganized manufacturing has been higher in the urban areas than that in the rural areas. The faster growth in the number of enterprises in the urban areas could be due to the change in the location of the enterprises which could be an outcome of both promising enterprises shifting actually to the urban areas and the reclassification of rural areas as urban over time.

An important point that comes out clearly from Tables 9 and 10 is that the growth rate in value added in unorganized manufacturing has been much faster than the growth rate in number of workers and number of enterprises during 1989–1990 to 2005–2006, which broadly corresponds with the reform period. The implication

Table 8 Growth rates of employment, real value added and number of enterprises in unorganized manufacturing, 1989–1990 to 2005–2006 (percent per annum)

Sector	Employment	Real value added	Number of enterprises
OAME	0.24	2.11	0.88
NDME	1.71	3.97	1.62
DME	1.32	5.77	1.48
Total unorganized manufacturing	0.65	3.96	0.97

Source Computed from NSS data on unorganized manufacturing enterprises (*OAME* own account manufacturing enterprises; *NDME* non-directory manufacturing establishments; *DME* directory manufacturing establishments)

Table 9 Growth in real value added in unorganized manufacturing, 1989–1990 to 2005–2006 (percent per annum)

Industries	OAME	NDME	DME	Unorg. Mfg
Food products, beverages and tobacco	0.83	1.91	7.25	2.58
Textiles and leather	6.81	6.07	6.57	6.53
Paper and products	6.44	0.26	2.10	2.00
Chemical and chemical products	3.98	−0.49	0.20	0.47
Non-metallic mineral products	1.60	7.04	8.30	5.99
Basic metals	6.30	1.33	5.36	4.81
Metal products	4.04	7.67	10.40	8.53
Machinery and equipment	1.30	2.65	8.32	5.48
Transport equipment	2.69	5.41	4.26	4.41
Other manufacturing incl. wood	−1.18	4.46	3.69	1.56
All industries	2.11	3.97	5.77	3.96

Source Computed from NSS data on unorganized manufacturing enterprises

Table 10 Employment growth in unorganized manufacturing, 1989–1990 to 2005–2006 (percent per annum)

Industries	OAME	NDME	DME	Unorg. Mfg
Food products, beverages and tobacco	0.94	−0.35	0.98	0.80
Textiles and leather	2.18	3.72	2.18	2.42
Paper and products	4.12	−0.93	0.76	1.43
Chemical and chemical products	3.26	0.17	1.12	2.06
Non-metallic mineral products	−2.24	−0.12	0.30	−1.21
Basic metals	3.07	1.23	−0.24	0.86
Metal products	1.32	3.71	2.04	2.26
Machinery and equipment	−3.19	−1.17	−0.37	−1.58
Transport equipment	−1.76	0.99	3.19	2.07
Other manufacturing incl. wood	−2.40	2.41	1.55	−1.35
All industries	0.24	1.71	1.32	0.65

Source Computed from NSS data on unorganized manufacturing enterprises

is that value added per worker as well as value added per enterprise has grown rapidly, particularly in the rural areas. However, the employment growth rate was extremely sluggish for which the productivity growth rate has been quite fast in the post-reform period and this needs to be interpreted carefully. Besides, the measurement of value added in the NSS surveys on unorganized manufacturing has possibly undergone major improvements over time and if so, the growth rates in value added are not strictly interpretable. Similarly, the definition of employment in these surveys to begin with has been quite loose and is not comparable with the NSS employment–unemployment survey, as mentioned above. Part of the decline in the employment growth rate over time in the unorganized manufacturing sector can be attributed to improvements in estimating the number of workers more rigourously.

Over the more recent period, i.e. 2005–2006 through 2010–2011, employment growth has been mostly negative in the own account enterprises (Table 11). However, in the establishments it was a little below 2 % per annum though across industry groups large variations are discernible. The aggregate employment figure for all establishments and own account enterprises turns out to be negative over 2005–2006 through 2010–2011.

One important structural change that has been occurring for a long time in the Indian manufacturing is the decline in the share of household enterprises: the share of household enterprises in manufacturing employment declined from 55 % in 1961 to 32 % in 1981 and further to 23.7 % in 1991(Ramaswamy 1994). This trend seems to have been arrested to a large extent in the post-reform period though not entirely. The household enterprises are covered under the OAME segment of the unorganized manufacturing. The share of OAME in unorganized manufacturing sector employment declined by 4.4 % points, from 69.4 to 65.0 % between 1989–1990 and 2005–2006. There was an increase in the share of NDME within unorganized

Table 11 Growth in employment and GVA in unorganized manufacturing: 2005–2006 through 2010–2011

Industry	OAME		Establishment		All	
	Workers	GVA in Rs. Crore (in 2004–2005 prices)	Workers	GVA in Rs. Crore (in 2004–2005 prices)	Workers	GVA in Rs. Crore (in 2004–2005 prices)
Food products, bev. and tobacco	−6.77	2.12	−3.39	−1.29	−5.94	0.35
Textiles and leather	2.48	15.14	0.46	9.84	1.79	12.07
Paper and products	−13.35	−4.04	15.53	23.45	−0.64	17.29
Chemical and chemical products	−17.36	−4.70	−2.04	4.59	−10.99	2.95
Non-metallic mineral products	−4.91	2.01	11.27	9.03	3.97	7.67
Basic metals	13.34	19.06	−2.30	−12.92	2.52	−9.10
Metal products	−3.69	9.65	4.77	1.47	1.82	2.54
Machinery and equipment	−26.19	−14.44	−8.65	−5.70	−12.64	−6.60
Transport equipment	−5.64	15.63	−7.70	4.20	−7.52	4.70
Other manufacturing incl. wood	−1.32	12.79	4.43	7.77	0.89	9.47
All industries	−2.45	9.58	1.85	5.25	−0.86	6.72

Source National Sample Survey Organisation

Table 12 Share of OAME, NDME and DME in employment and value added in unorganized manufacturing, 1989–1990 and 2005–2006 (percent)

Sector	Employment		Value added	
	1989–90	2005–06	1989–90	2005–06
OAME	69.4	65.0	42.7	32.0
NDME	13.4	15.9	24.1	24.1
DME	17.2	19.1	33.3	43.8
Unorg. Mfg.	100.0	100.0	100.0	100.0

Source Computed from NSS data on unorganized manufacturing enterprises

manufacturing in terms of employment (Table 12). On the other hand, the share of DMEs in terms of both employment and value added rose over the same period. Two broad conclusions have been drawn by looking at the data for rural and urban areas separately (Goldar et al. 2011): (1) In rural areas, the share of OAME in value added declined in general, between 1989–1990 and 2005–2006, except paper and chemicals. (2) In urban areas, the share of DME in value added has generally improved except non-metallic minerals and transport equipment.

4.2 Growth in Labour Productivity

Growth in labour productivity (defined as real gross value added per worker) turns out to be positive in all the three OAME, NDME and DME sectors (1989–90 to 2005–06, Table 13). In both NDME and DME, chemical and chemical products recorded a negative growth in labour productivity in the nineties. As a result, the combined unorganized manufacturing growth rate in this industry group turned out to be negative. In industries like basic metals, transport equipment and other manufacturing productivity growth has been above average. Over the recent years (2005–2006 to 2010–2011), the labour productivity growth has been very fast in the unorganized manufacturing (i.e. an almost of 8 % per annum) since the employment growth has been almost negative. Usually in the unorganized manufacturing, the decline in employment is expected to be accompanied by a proportionate decline in value added as labour intensive technology is used extensively. However, this does not seem to have happened. Is it because the redundant labour has been retrenched without causing any negative effect on value added or is it because labour-saving technology is being pursued even in the unorganized component? From Table 10, we noted that the value added growth even in the OAMEs which are believed to be more vulnerable compared to the establishments has been substantial. If these units are not expected to have access to modern technology, the retrenchment of redundant labour and/or improvement in the definition of worker used in the survey must be the cause of decline in employment growth which has not been accompanied by decline in value added growth. However, there could be a third line of argument as well: the unorganized manufacturing units now cater to many

Table 13 Labour productivity growth in unorganized manufacturing, (percent per annum)

Industries	Unorganized manufacturing				
	1989–1990 to 2005–2006				2005–2006 to 2010–2011
	OAME	NDME	DME	ALL	ALL
Food products, beverages and tobacco	−0.11	2.26	6.21	1.77	6.68
Textiles and leather	4.53	2.26	4.29	4.01	10.11
Paper and products	2.23	1.20	1.33	0.55	18.04
Chemical and chemical products	0.70	−0.66	−0.91	−1.56	15.66
Non-metallic mineral products	3.93	7.16	7.97	7.30	3.56
Basic metals	3.13	0.10	5.61	3.92	−11.33
Metal products	2.69	3.81	8.20	6.13	0.71
Machinery and equipment	4.63	3.86	8.72	7.17	6.92
Transport equipment	4.52	4.37	1.04	2.29	13.21
Other manufacturing incl. wood	1.25	2.01	2.12	2.95	8.51
All industries	1.87	2.21	4.39	3.38	7.65

Source NSS data

organized sector units through contractors. And from our micro surveys we have observed that work consignments are large in number but at the same time the large units are particular in terms of quality-specificity. In other words, the unorganized sector enterprises are now conscious enough to manufacture quality products for which they have to appoint relatively better quality labour with higher productivity in the place of too many unskilled workers. Also, the fact that the payment system is now more of piece-rate type has motivated the enterprises to become more competitive by appointing better performing workers.

Wages or emoluments per worker grew at a fast rate in the unorganized manufacturing sector (Table 14). It may be useful to examine the wage–productivity relationship in order to reflect on the relative efficiency of workers: wage growth vis-à-vis productivity growth. Wage elasticity with respect to productivity, which is defined as the ratio of the rate of growth in wages to the rate of growth in productivity calculated over the period 1989–1990 through 2005–2006 is extremely high in the unorganized manufacturing at the aggregate level (Table 15). Even across various groups of industries it turns out to be unity or more than unity, implying wage growth has been more than the productivity growth. However, we need to recognize the fact that the base year wage in the unorganized sector has been much lower than that in the organized sector and, therefore, a higher growth rate corresponding to the unorganized manufacturing sector does not come as a surprise. Notwithstanding these explanations, we may note that globalization has resulted in a higher degree of mobility across activities and regions and in an

Table 14 Growth rate in wages per worker in unorganized manufacturing, 1989–1990 to 2005–2006 (percent per annum)

Industries	Unorganized manufacturing		
	NDME	DME	Combined
Food products, beverages and tobacco	6.91	7.98	7.48
Textiles and leather	6.31	6.26	6.33
Paper and products	4.54	4.30	4.86
Chemical and chemical products	4.58	1.80	2.52
Non-metallic mineral products	11.18	7.08	7.72
Basic metals	4.49	4.28	4.38
Metal products	8.35	10.43	9.85
Machinery and equipment	8.87	6.52	7.65
Transport equipment	7.57	3.96	4.90
Other manufacturing incl. wood	4.41	2.75	3.50
All industries	6.19	5.48	5.88

Source NSS data

Table 15 Elasticity of wages with respect to productivity in unorganized manufacturing (1989–1990 to 2005–2006)

Industries	Unorganized manufacturing		
	Gr. rate Wages (% p.a.)	Gr. rate Productivity (% p.a.)[a]	Elasticity
Food products, beverages and tobacco	7.48	4.32	1.73
Textiles and leather	6.33	3.37	1.87
Paper and products	4.86	1.50	3.23
Chemical and chemical products	2.52	−0.81	−3.10
Non-metallic mineral products	7.72	7.83	0.99
Basic metals	4.38	4.43	0.99
Metal products	9.85	6.29	1.57
Machinery and equipment	7.65	7.13	1.07
Transport equipment	4.90	1.86	2.64
Other manufacturing incl. wood	3.50	2.05	1.71
All industries	5.88	3.52	1.67

[a]For NDM and DME only, to be consistent with the estimated growth rate in wages
Source NSS data

attempt to prevent high labour turnover cost it is natural for the employers to offer higher wages. Hence, part of the rapid growth in wages in the unorganized sector is in response to the changing economic environment. Also, subcontracting from large units is believed to have raised the supply of work consignments in the unorganized sector which possibly contributed to rapid wage growth. Subcontracting from the large units requires quality control which can be assured partly through improving the tenure of the workers, i.e. by reducing the labour turnover rate and introducing regularity in the informal/unorganized sector. This is possible only through wage increase.

5 Inter-Industry Linkages

In order to understand the inter-industry input-output linkages within the manufacturing sector, we have considered the Social Accounting Matrix (SAM) prepared by Pradhan et al. (2013). The SAM for 2007–2008 has 49 sectors and is based on input-output I-O table 2006–2007. From this, we have extracted the manufacturing component which has 14 sub-sectors in total. In this study, we focus only on the intra sub-sectoral dependence within the manufacturing sector and not the inter-sectoral relationships such as agriculture and manufacturing or manufacturing and services. Though there is a great deal of evidence on growing inter-linkages between manufacturing and services (Banga and Goldar 2004; Dasgupta and Singh 2005; Mitra and Schmid 2008), this study confines to the manufacturing sector only. Kant (2013), in fact, presents an estimate of the multiplier effect of every job created in the manufacturing sector, which is around three jobs created in the tertiary/services sector.

The column total in Table A.10 adds up to 1 indicating the inputs drawn from various sub-sectors (the same and others) for the production of one unit of output of a given sub-sector. On the other hand, in Table A.11 the row total adds up to one giving the distribution of one unit of output of a given sub-sector across various sub-sectors used as inputs. It may be noted that inputs to and inputs from sectors other than manufacturing are excluded in the analysis. The findings are as follows.

In deciding whether the input drawn from a sub-sector is sizeable or not, we have taken 5 % as the benchmark. Petroleum and chemicals are some of the key industries which provide inputs to the production of goods in most of the industries. Particularly, the consumer goods and light goods industries such as food, beverages, textiles, wood, paper, leather etc. are mostly dependent on these products. However, as we move on to other groups such as fertilizer, chemical, cement, metal and machinery, we observe that their dependence on each other is quite significant in addition to the two key inputs, petroleum and chemicals. The transport goods sub-sector for example, depends on a number of sub-sectors such as wood, paper, rubber, chemical, metal, metal products, non-electrical machinery and electrical machinery (Table 16). This may tend to suggest that any deceleration in the labour-intensive light goods sectors does not necessarily impact on the heavy goods sector adversely.

On the other hand, from Table A.11 presenting the distribution of one unit of output of a specific sub-sector across various sub-sectors in terms of input-use, we observe that items such as food, beverages are used as inputs in those very sub-sectors. However, in rest of the cases, the utilization of products across sub-groups other than the same one is quite distinct. Hence, we may conclude that

Table 16 Major inputs used

Industry	Depends on
Food depends on	Petroleum, Fertilizers, Chemicals
Beverages depends on	Petroleum, Fertilizers, Chemicals
Textiles depends on	Petroleum, Fertilizers, chemicals
Other textiles depends on	Petroleum, Fertilizers
Wood depends on	Petroleum, Fertilizers
Paper depends on	Petroleum, Fertilizers, Chemicals
Leather depends on	Food, Chemicals
Rubber depends on	Rubber, Petroleum, Non-electrical, Machinery, trans other
Petroleum depends on	Other, Textile Wood, Petroleum, Transport
Fertilizers depends on	Chemicals
Cement depends on	Rubber, Petroleum, Chemical, Metal
Non-metals depends on	Petroleum, Chemical, Metals, Non-electrical
Metals depends on	Food, Paper, Petroleum, Chemicals, Non-electrical
Metal products depends on	Food, Beverage, Paper, Rubber, Chemical
Non-electrical depends on	Textile, Petroleum, Chemical
Electrical machinery depends on	Textile, Other textile, Chemical, Non-electrical
Transport depends on	Wood, Paper, Rubber, Chemical, Metal, Metal products, Non-electrical, Electrical
Other Mfg depends on	Paper, Chemical
Fertilizers depends on	Petroleum, Chemical, Metal products, Non-electrical, Machinery, Transport
Chemicals depends on	Petroleum, Chemical, Non-metal, Metal, Non-electrical

Source See Table A.10

Table 17 Sector of use as inputs

Sector	is used in
Food is used in	Food, Beverage
Beverage is used in	Beverage, Chemical
Tex is used in	Textile, Other textile
Other tex is used in	Textile, Other textile, Chemicals, Electrical
Wood is used in	Food, Paper, Chemicals, Non-electrical, Electrical
Paper is used in	Food, Paper, Chemical, Electrical
Leather is used in	Other textile, Leather, Electrical, Transport
Rubber is used in	Beverage, Rubber, Chemicals, Non-electrical, Electrical, Transport
Petroleum is used in	Food, Petroleum, Fertilizers, Chemicals, Non-metal, Metal
Other Mfg is used in	Electrical, Other
Fertilizers is used in	Chemicals
Cement is used in	Chemicals, Non-metal
Non-metals is used in	Cement, Non-metal, Metal, Electrical, Non-electrical
Metals is used in	Metal, Metal-products, Electrical, Transport
Metal products is used in	Metal, Metal-products, Non-electrical, Electrical, Transport
Non-electrical is used in	Other textile, Non-electrical, Electrical, Transport
Elect machinary is used in	Non-electrical, Electrical, Transport
Transports is used in	Metal, Transport
Food, Fertilizers, Chemical	
Chemicals is used in	Rubber, Chemicals, Electrical

Source See Table A.11

some of the labour-intensive industries do not have strong inter-industry linkages whereas some others like textile etc. do have inter-connections, implying that industrial deceleration in the heavy goods sector can reduce the input demanded from the labour as well as capital-intensive sub-sectors (Table 17). Thus, the growth and employment in the labour-intensive sub-sectors may suffer which in turn may affect adversely the overall employment growth in the manufacturing sector. This may also have cascading effects on the rest of the economy and the pace of employment generation and the effectiveness of the industrial sector in reducing poverty.

6 Productivity, Skills and Innovation

Skill formation is an indispensable prerequisite for labour productivity as well as total factor productivity growth. Even if jobs can be generated in the process of growth they may require certain skill and knowledge which the available labour may not provide. Hajela (2012) argues that a shortage of skills is making more people unemployable in India. There are 17 central government ministries that offer skill development initiatives through school education, institutes of higher learning and specialized vocational training institutes. China, with a similar scale of population and training structure to that of India, has higher labour productivity, indicating higher skills. She argues that India lacks sufficient skilled workers as its existing vocational training system does not target the casual or informal workforce, who constitutes a sizable percentage of India's working population. Froumin et al. (2007) urged that only 16 % of Indian manufacturing firms offer in-service training, compared with 92 % in China and 42 % in the Republic of Korea. The Indian firms that provide in-service training are 23–28 % more productive than those that do not.

The National Skill Development Corporation (NSDC) in India was formed to achieve the target of skilling/upskilling 150 million people by 2022 by fostering private sector initiatives in the skill development space. To support its various initiatives, NSDC is looking at creating an enabling environment by developing a robust research base for skilling. It conducts studies to understand the geographical and sector-wise skill requirements and on various subjects that can influence and enable the skilling environment in India. For a wide range of sectors such as auto and auto component, banking, tourism, infrastructure and also the unorganized sector NSDC reports are prepared. Several studies across countries highlight the mismatch between skill requirement and the quality of labour available. Richardson (2007) suggests the following scheme for classifying skills shortages:

- Level 1 shortage: there are few people who have the essential technical skills who are not already using them and there is a long training time to develop the skills.
- Level 2 shortage: there are few people who have the essential technical skills who are not already using them but there is a short training time to develop the skills.

- Skills mismatch: there are sufficient people who have the essential technical skills who are not already using them, but they are not willing to apply for the vacancies under current conditions.
- Quality gap: there are sufficient people with the essential technical skills who are not already using them and who are willing to apply for the vacancies, but they lack some qualities that employers consider are important.

Comyn (2012) reports on recent research into enterprise skill profiles and workplace training practices in the Bangladesh manufacturing industry. Based on data from 37 enterprises across eight manufacturing groups, collected during a study for the International Labour Organisation, he analyzed enterprise and sectoral skill intensity and identified key skill issues. This helps prioritize sectors for project-based investments in workplace training and industry skill development. Particularly, at a time of significant expansion negligence of research and training can affect performance adversely. The research also illustrates the difficulties of using generalized approaches to classifying and comparing skills at the enterprise and sectoral levels. Whilst the concept of skill intensity and the use of occupational classifications may be interesting it does not talk about the relevance of the acquired skill in a particular occupation across other occupations. Without a skill which has greater applicability across a number of sectors, the bargaining power of the workers and consequently the occupational mobility tend to change sluggishly.

Desjardins and Rubenson (2011) throw light on the potential causes of skill mismatch, the extent of skill mismatch, the socio-demographic make-up of skill mismatch, and the consequences of skill mismatch in terms of earnings as well as employer-sponsored adult education/training. The authors have made a distinction between *skill mismatch* and *education mismatch*. Two key findings are as follows. First, including supply and demand characteristics in an earning function it reveals that labour demand characteristics are more important than labour supply characteristics in explaining earnings differentials. In other words, skills matter for earnings but only if they are required by the job. Second, the skill content of jobs seems to be an even stronger determinant of participation in employer supported adult education/training than educational attainment or literacy proficiency. On the whole, they argue that skill formation is not just a supply side issue; it is just as much a function of work tasks and work organization on the demand side. Policies on skill formation have to take into account both the supply and the demand side. Identifying the mechanisms that help to foster the optimal utilization of the existing skill base is essential. Otherwise, many workers even with high qualifications may lose their skills due to a lack of use, which is an erosion of educational investments.

In the context of small versus large firms and skill acquisition, Bishop (2012) points out that due to certain characteristics inherent to their small size, small firms generally display greater informality in their learning processes. They cannot normally be expected to learn in the highly formalized and structured ways more often pursued by their larger counterparts. But small firms can and do benefit from formal training also. Bishop (2012) deals with the concept of "learning architecture" to illuminate the connection between firm size and learning processes.

Education level to determine skills is a fairly blunt instrument, more so given that the notion of 'skill' is socially constructed, but that's another argument and linked to the new National Skills Qualification Framework (NSQF). As far as the question of which skills are important, the argument about the issue of skill shortages and the importance of expansion of higher education system should go hand in hand. Higher education may bring higher returns on investments only if the person is employed. Also, the analysis of employability of those with vocational type training may bring a different insight into the story of lower returns on investments into vocational training, if the employability is high. Several studies have used education as a base. Tilak (2003) based on the evidence in Asia and the Pacific countries observed significant effects of higher education on development. In terms of NSS data, Mehrotra et al. (2013) uses the educational (general, technical and vocational) attainments to understand skill levels of the existing workforce. They estimate the skilling requirements, sector-wise, under different scenarios to arrive at a realistic and desirable target and find that the challenge of skill development—both in quantitative and qualitative terms—is enormous and requires a careful policy stance. Mitra et al. (2002) noted that in most of the two-digit industry groups of the manufacturing sector education, health and other indicators of the social infrastructure impacted on the total factor productivity growth much more than the physical or the financial infrastructure. Hence, the quality of human capital is a strong determinant of performance. In the present study on Indian manufacturing after bringing out certain aspects of sluggish growth in demand for labour, it is important to assess the employability of the existing labour force. There is a view that highlights the supply side factor pointing to the poor human capital formation, reducing the performance indicator which in turn motivates employers to adopt capital-intensive technology. On the other hand, given the complementary relationship between capital and skill capital-intensive technology does not necessarily assure optimal utilization of the available technology due to skill shortage. Mitra (2009) noted that imported technology which is capital intensive in nature tends to reduce the technical efficiency in the manufacturing sector in the developing countries, which could be due to the unavailability of skill.

Based on the NSS data, we have estimated the index of skill mismatch following the methodology of Estevão and Tsounta (2011).

$$\text{Skill Mismatch Index} = \text{sigma } (j = 1 \ldots n) \left(S_{jt} - M_{jt} \right)^2$$

where, j is skill level, n is the number of skill categories, S_{ij} proportion of population with skill level j at time t and M_{ij} is proportion of employees with skill level j at time t.

We have also estimated a similar index taking the difference in the proportional distribution of working persons and the unemployed across various skill levels.

A third set is estimated by taking into account the proportional distribution across various skill levels of workers in each of the activities and also all workers between the following two time points: 2004–2005 and 2009–2010.

6.1 Results on Skill Mismatch Index

The index representing the difference between the skill level of the population and the workers is estimated at 73.11 (based on Table 18), which is quite high. In other words, the difference between the skill levels of the potential labour supply and those already working is sizeable. Of course, we need to understand that those who are employed are not necessarily absorbed in demand-induced activities. There are several activities which are repository of surplus labour, not requiring much skill to be pursued. Hence, those who are working are not necessarily better off compared to the non-workers in terms of skill levels. Even then the mismatch is significant.

The inter-temporal skill mismatch index as estimated from the distribution of workers in each of the activities across various skill levels is again highly significant (Table 19). The skill gap in the manufacturing sector is seen to be on the high side indicating that over time jobs in this activity are becoming more skill-based compared to those in other activities such as construction and transport.

Even a cursory look at the percentage of population with education shows that only around one-fifth of the population/workforce had studied either up to higher secondary or more in the age brackets of 30–44 and 45–59, in 2004–2005. More than half of the children in the age group of 7–14 had not studied at least up to primary level. All this shows poor employability of the available labour supply (Table 20). However, there has been a rise over time in the percentage of the population/work force who acquired the specified level of education as seen from the 2009–2010 survey data.

Table 21 list the percentage of population and workforce with no education, up to primary level and those with above primary and those with technical education. All the three categories may add up to more than 100 % as some individuals may have acquired both general and technical education. These tables also identify only one third of the population and a little below half of the work force who can be considered to be relatively better in terms of skill endowment, in the year 2004–2005. The corresponding figures, however, went up to a little above two-fifth of the population and a little above half of the work force in 2009–2010 (Table 21).

If we consider those who acquired vocational training formally or informally or through self-learning—the information for whom is given only for the age cohort 15–29—their percentage share in the total population in the corresponding age bracket turns out to be only 11 and 7 for 2004–2005 and 2009–2010 respectively. However, if we consider the set of workers in this age bracket, the respective figures go up to 15.6 and 10 % for both the years.

6.2 Wage Function

Another important way of showing the importance of skill in the job market is to calculate the returns to skill and the differences in the returns across various levels

Table 18 Percentage distribution across educational levels: 2009–2010

Category	Not literate	Up to primary	Middle	Secondary	Higher secondary	Diploma/certificate course	Graduate	Post-graduate	All
Workers	31.7	24.8	16.9	11.5	6.1	1.2	5.8	1.9	100
Unemployed	5.3	12.5	17.3	15.4	16.2	6.2	20.4	6.7	100
Population	33.2	32	13.8	9.3	5.7	0.8	3.9	1.1	100

Source Based on NSS Data, 2009–10

Table 19 Skill mismatch index across activities over 2004–2005 to 2009–2010

Activities	Manufacturing	Electricity	Construction	Trade	Transport	Financing	Services	Total workers
Index	48.86	73.88	21.41	53.21	21.41	46.12	46.82	88.38

Note All workers include agriculture and mining also
Source Based on NSS Data, 2004–05 and 2009–10

Table 20 **a** Education level: total population. **b** Education level: work force

Year	Up to 6	7–14 (% primary and above)	15–29 (% secondary and above)	30–44 (% higher secondary and above)	45–59 (% higher secondary and above)	>59 (% higher secondary and above)	Total (% above the threshold limit)
(a)							
2004–5	Omitted	42.06	38.54	19.68	15.45	6.67	28.9
2009–10	Omitted	47.88	50.63	25.16	18.48	10.61	35.4
(b)							
2004–5	Negligible sample	38.24	31.04	21.78	18.32	6.92	23.25
2009–10	Negligible sample	48.34	41.68	28.24	22.50	10.14	29.59

Source Based on NSS Data, 2004–05 and 2009–10

Table 21 **a** General and technical education: total population. **b** General and technical education: work force

	Type	Up to 6	7–14	15–29	30–44	45–59	>59	Total
(a)								
2004–2005	Illiterate	79.87	9.43	16.66	32.34	42.14	60.62	34.21
	Up to primary	20.12	79.44	22.60	24.91	24.38	22.13	33.37
	Above primary and also those with technical education	0.01	11.21	63.61	45.98	35.97	18.40	34.24
2009–2010	Illiterate	75.11	4.79	10.54	24.17	34.96	53.63	27.43
	Up to primary	24.89	81.09	17.22	21.33	22.94	21.24	31.15
	Above primary and also those with technical education	0.00	14.12	74.99	57.10	43.96	26.20	43.04
(b)								
2004–2005	Illiterate	Negligible sample	40.13	21.04	31.23	39.43	55.32	31.52
	Up to primary		49.84	27.65	24.66	24.29	25.64	25.99
	Above primary and also those with technical education		10.09	54.59	47.99	39.44	20.38	45.78
2009–2010	Illiterate	Negligible sample	29.08	13.66	21.67	30.87	47.43	23.50
	Up to primary		56.83	23.34	21.14	22.84	25.31	22.77
	Above primary and also those with technical education		13.36	66.38	60.61	48.87	28.53	56.75

Note All the three categories may add up to more than 100 % as some individuals may have acquired both general and technical education
Source Based on NSS Data, 2004–05 and 2009–10

of skills. One of the convenient ways of conceptualizing the returns is to estimate an earning function with dummies representing different educational levels. The weekly wage function estimated on the basis of the NSS unit level data from the 66th round shows that education dummies (the general and technical both) tend to enhance earnings (Table 22). Those who have general education tend to have higher earnings in comparison to the illiterates. In fact most of the education dummies are significant except Gedu2 which represents literates without formal schooling. Similarly technical education dummies are also significant. But those who have acquired vocational training are not better off in comparison to those who do not have it. Male earnings are higher than the female earnings. Scheduled castes receive a lower wage compared to the general-cum-OBCs. Large households are also worse off in terms of earnings, suggesting lower productivity levels and poor quality of labour. Though age, taken to be a proxy for job market experience, is not significant, age square is and it has a positive coefficient. All this tends to suggest that skill

Table 22 Dependent variable: regression equation for weekly wage (in Rs per person)

	Coefficient	T ratio
Age	−1.98	−0.58
Agesq	0.39	8.98*
hhhold_size	−10.28	−4.57*
Gedu2	18.63	0.24
Gedu3	201.06	13.44*
Gedu4	831.28	46.45*
Gedu5	2028.29	95.56*
Tedu2	2756.52	47.92*
Tedu3	690.78	20.12*
Tedu4	903.45	19.08*
vtrn1	−766.49	−24.02*
vtrn2	−124.35	−3.89*
vtrn3	−77.98	−3.59*
mst1	−8.88	−0.29
mst2	193.45	7.68*
gender1	371.49	27.14*
reldum1	−184.97	−10.26*
reldum2	−155.69	−6.42*
Socdum1	26.93	1.56
Socdum2	−101.55	−7.4*
occ1	−700.31	−16.03*
occ3	−412.47	−9.3*
occ4	−586.32	−13.42*
occ5	−750.70	−16.42*
occ6	−323.05	−7.09*
occ7	101.45	2.34*
Secdum1	−258.73	−22.27*

Note Number of obs = 75518, F(27, 75490) = 1914.54, Adj R-square = 0.4062

Age Age of the earner, *Agesq* age square, *hhhold_size* household size, *Gedu2* literate without formal schooling, *Gedu3* primary and middle level, *Gedu4* secondary and higher secondary, *Gedu5* graduation and post graduation, *Tedu2* technical degree in agriculture/engineering/technology/medicine, etc., *Tedu3* diploma or certificate (below graduate level) in: agriculture, engineering/technology, medicine, crafts, other subjects, *Tedu4* diploma or certificate (graduate and above level) in: agriculture, engineering/technology, medicine, crafts, other subjects, *vtrn1* receiving formal vocational training, *vtrn2* received vocational training: formal, *vtrn3* received non-formal: hereditary, self-learning, learning on the job, others; *mst1* never married, *mst2* currently married, *gender1* male, *reldum1* Hindu, *reldum2* Islam, Socdum1: schedule tribe, *Socdum2* schedule caste, *occ1* agriculture, hunting, forestry, fishing, *occ3* manufacturing, electricity, gas and water supply, *occ4* construction, *occ5* wholesale and retail trade, hotel and restaurant, *occ6* transport, storage and communication, *occ7* other services including financing etc., *secdum1* rural dummy

Source Based on NSS unit level data (66th Round, 2009–10)

*Represents significance at 1 per cent level

formation does contribute to higher levels of earnings. Though ideally it would have been desirable to estimate an earning function with both demand and supply side variables, due to unavailability of data on the demand side such an equation could not be pursued. The sector-specific dummies are mostly negative suggesting a lower earning compared to mining and quarrying. However, the occ7 representing other services including financing etc. has a positive coefficient indicating higher earnings in this sector. In relation to the manufacturing sector, we may infer that those who are able to acquire skill are able to secure employment with higher levels of earnings as demand exists for the skilled variety of labour due to rise in skill intensity of the technology.

7 Innovation and Industrial Employment

7.1 Various Viewpoints: Existing Studies

Schumpeter (1939, 1961) initiated the concept of "innovation". In his postulation innovation is a new production function, displaying a new combination of factors of production or production conditions. Innovation is a continuous process of creative destruction, old being replaced by the new. The combination of capital, labour and other factors of production is optimized in the process of innovation and its impact on total employment and employment structure is cyclical. In the initial stages, total employment grows sluggishly or even declines, while employment structure does not change significantly; at a later stage, there are rapid increases in total employment and marked changes in employment structure; and in the final stage of innovation, changes both in total employment and employment structure gradually diminish until the next innovation comes through (Guangrong and Yuanyuan 2009).

If the new technology enhances productivity as well as promotes employment, the choice is clear. Such a possibility, though empirically difficult to materialize, exists at least theoretically. For example, technological progress brings in upward shift in the production frontier, which would mean higher levels of output for the given levels of inputs. In such a situation if the new technology becomes labour intensive, the rise in value added and employment both will be witnessed. However, the value added growth will be more than the rise in employment, and hence, labour productivity can actually shoot up.[29] Conversely, the new technology can dampen employment and improve productivity by adopting capital-deepening process.[30]

---

[29]However, when output is fixed, the shift in technology from being capital intensive to labour intensive would result in deterioration in labour productivity.

[30]Materials for this chapter have been drawn from Mitra and Jha (2015).

7.1.1 Negative Relation Between Innovation and Employment

Choi et al. (2002) analyzed the implications of Hicks-neutral technical progress for a small Harris-Todaro economy with variable returns to scale. The analysis demonstrates that the welfare effects of technical progress consist of three components, i.e. the primary growth effect, the returns-to-scale effect and the employment effect. This type of decomposition is indeed useful as it deciphers the effects of technical progress into various components. Besides, the study works out the possibilities under non-constant returns to scale which is a much stronger possibility in the real world than a constant returns to scale situation. Under constant returns to scale, the possibility of poverty-aggravation may not exist and one may conclude that technical progress will be beneficial. But with the introduction of non-constant returns to scale, technical progress can lead to the returns-to-scale effect, which can be of any sign, and the sum of the primary growth effect and the employment effect again can be of any sign. In other words, growth without employment generation is possible as technical progress tends to reduce labour absorption.

Technical progress and rising capital intensity in the literature are almost synonymous. On the other hand, innovation in the line of labour intensive technical progress is a difficult proposition. The capital-intensive technical change also has important implications for rates of industrialization and capital accumulation even when the economies, particularly in the developing world, are characterized by a dual economic structure. Kelley et al. (1972) noted that increases in the bias may tend to inhibit the rate of industrialization and reduce the rate of capital accumulation without appreciable changes in per capita GNP growth. Related to these results is the extent to which labour absorption in the industrial sector is affected. The study observed an important retarding influence that accumulates over time. It questions the wisdom of introducing labour-saving technology in the industrial sector in order to enhance per capita growth. The authors rather noted that per capita income is mostly insensitive to the technological bias introduced in the industrial sector of the developing countries. Hence, the outcome is neither an increase in per capital income nor a rise in employment in the industrial sector in response to adoption of capital-intensive technology.

In fact, Mureithi (1974) elucidates this point with great lucidity. The rising capital–labour ratio means that each job creation becomes more capital-expensive. Of course it must not be supposed that rising capital intensity is bad per se as a large part of the capital formation could be devoted to the building of infrastructure like roads, public works, communications, etc.

7.1.2 Positive Association Between Innovation and Employment

In addition, as Mureithi (1974) argues, it is pertinent to realize that production actually takes place in stages: (1) material handling, (2) material processing, (3) material handling among processes, (4) packaging, (5) storage of the finished products. Of the five stages, only the second, i.e. the central processing, is capital intensive because at this stage the finer precision of temperature, pressure, ingredients combination, etc., is important. But there are many other stages where factor substitutability is technically possible and thus the entrepreneurs have a choice to select the technology. The desirability of a technology has to be judged not merely by its scientific or technical sophistication, but rather by its appropriateness in the context of the society in which it will be used. It requires innovative ideas to reduce the labour-saving elements of a technology while maintaining or improving quality and efficiency. Even after accounting for the fact that there could be stages where capital-intensive technology is absolutely necessary, innovation and employment can move in a positive direction in many other stages which then can offset the negative association between technology and employment as conceived in certain specific stages.

The "compensation theory" as Vivarelli (2011, 2013) points out, argues that technological unemployment is a temporary phenomenon. The labour-saving effects of technology can be offset through: "(1) additional employment in the capital goods sector where new machines are being produced, (2) decreases in prices resulting from lower production costs on account of technological innovations, (3) new investments made using extra profits due to technological change, (4) decreases in wages resulting from price adjustment mechanisms and leading to higher levels of employment, (5) increases in income resulting from redistribution of gains from innovation and (6) new products created using new technologies" (Vivarelli 2013).[31]

Another interesting point in relation to the preference for new technology is as follows (James 1993). If new technology is not adopted it may affect the quality of products as well as exports, resulting in employment loss. On the other hand, adoption of new technology which is capital intensive in nature can cause employment to fall. Hence, one has to compare between employment loss due to drop in exports prompted by the traditional labour-intensive technology and employment loss due to adoption of capital-intensive technology to assess which one is greater in magnitude. Further, the speed of production, product flexibility and location-specific factors need to be considered in assessing the total effect of technology on employment. If certain products are manufactured in the low-cost countries labour-intensive technology can still be pursued. Hence, the factor price ratio is an important determinant of technology choice and decision about location of production base, which eventually impact on employment. The idea of enlarging the production base across the globe is embedded in the study by James (1993). While the low labour cost countries can specialize in the production of certain goods or certain components of the composite goods using the labour intensive methods, the developed countries may specialize in certain other components that require very high levels of capital and skill. Thus, the newer and innovative ways would mean that technical progress would not only suit the labour market situation

---

[31]Also see Vivarelli (1995) and Pianta (2005).

of the developing and the developed countries both but also bring in a positive relationship between innovation and employment at large.

Bogliacino (2014) using company data from R&D Scoreboard for Europe analyzed the microeconomic relationship between innovation and employment. He observed the prevalence of scale effect for a given R&D intensity generating an increasing relationship between total turnover and employment. Bogliacino and Vivarelli (2012) using database of 25 manufacturing and service sectors across 16 European countries and applying GMM-SYS panel estimations of a demand-for-labour equation augmented with technology noted that R&D expenditures have a job-creating effect. Bogliacino et al. (2012) tested the job creation effect of business R&D applying the dynamic LSDVC estimator to a longitudinal database of around 677 European companies and found it to be positive in services and high-tech manufacturing though not in traditional sectors. Using the data from the three surveys on Italian manufacturing firms, Hall et al. (2008) found no evidence of significant employment displacement effects arising from process innovation. Though Harrison et al. (2014) using comparable firm-level data from France, Germany, Spain and the UK detected large displacement effects induced by productivity growth in the production of old products, the effects related to product innovations were seen to be strong enough to more than compensate these displacement effects.

A positive relationship between innovation and employment has been conceptualized in a novel way by Saviotti and Pyka (2004). Interpreting economic development as synonym for new goods, services or sectors they view it as a result of increasingly systematic use of innovation. It is quite natural that as the old product or services matures employability declines. Thus, to improve the level of employment in a continuous manner, innovation has to go on and new goods and services have to be produced. In this sense, innovation and employment can go hand in hand. The ability to reap variety is a manifestation of economic development, which in turn can create employment steadily. Also, on the productivity front its growth may not take place indefinitely implying upper bounds on sectoral productivity growth. In order to augment the productivity growth at the country-level, efforts have to be pursued to create new sectors. On the whole, the possibility of a positive relationship between innovation, employment and growth is very much comprehensible.

7.1.3   Innovation and Type of Labour

Next, one may pose the question in relation to product and process innovation. The interaction between economic integration, product and process innovation, and relative skill demand is an important aspect, which Braun (2008) analyzes in a model of international oligopoly. Lowering of trade barriers increases the degree of foreign competition which may have effects on the incentives of firms to undertake R&D investment and also the firms' demand for skilled relative to unskilled workers. Increased competition following economic integration induces firms to bring down production costs by investing more aggressively in process R&D. At the same time, competitors expand their investments in product innovation in order to reduce the substitutability of their products. However, all this would require highly skilled human labour which can initiate newer ways of introducing cost-efficient production processes and bring down the product differentials between the imported goods and the domestically produced goods. On the whole, economic integration and innovation are interlinked resulting in an increase in the relative demand for skilled workers[32] and not the unskilled or semi-skilled variety of labour force which is in excess supply in most of the developing countries. Innovation and skill intensity usually go together—hence, even if innovation is not always labour displacing it benefits only those who are relatively in short supply. This tends to indicate that wage inequality is likely to increase in the process of innovation and increased trade.

On the empirical front Berman and Machin (2004) showed the skill-bias of technological change especially in middle-income countries. Pianta (2005) emphasizes that innovation-based growth and job creation may operate in drastically different ways during different phases of the cycle, implying that the employment dynamics are not affected by the same factors and in the same ways during the upswings and the downswings. Piva (2003) presents a critical comparison of the positive implications of technology transfers (such as positive spillovers, technological catching-up, growing complementarities with domestic firms) with the negative ones (displacement of workers, negative welfare implications, competitive effects with domestic firms). Also, the author considers the nature of transferred technologies (labour-saving and/or skill-bias, embodied or not embodied in capital), together with the different institutional 'absorptive capacities' and sectoral specializations of both middle-income and low-income developing countries.

Lee and Vivarelli (2006) suggest that import of capital goods may imply an increase in inequality via skill-biased technological change. Imports of capital goods,—embodying technological innovations—are important because of the role they play in contributing to capital upgrading and more generally to the economic growth of the developing countries. In fact, even without necessarily assuming that developed countries transfer their "best" technologies, transferred technologies are relatively skill-intensive, i.e. more skill-intensive than those in use domestically before trade and FDI liberalization. Thus openness—via technology—should imply a counter-effect to the SS theorem prediction, namely an increase in the demand for skilled labour, an increase in wage dispersion and so an increase in income inequality.

Castellani and Zanfei (2006) present an in-depth theoretical and empirical analysis of the key issues underpinning the relationship between innovation and multinational companies. The authors argue that neither every foreign firm is a good source of externality nor every domestic firm is equally well placed to benefit from

---

[32]Vivarelli (2011) argues that innovation has a strong skill-bias.

multinationals. Spillovers from multinationals differ according to the technological profiles, embedded-ness and linkage creation of both foreign and domestic firms. Hasan (2002) presented evidence from panel data on Indian manufacturing firms in favour of a significant effect of imported technology on productivity. In general, the empirical literature on R&D, using cross-sectional data, reports strong evidence in favour of its positive effect on productivity while the time series estimates are less conclusive (Crespi and Pianta 2006). Using data on 33 Indian manufacturing industries in India for the period, 1992 through 2001, Pandit and Siddharthan (2006) further showed that technology imports, through joint ventures and MNE participation, influence employment positively. They noted that employment growth, production of differentiated products, skill intensity of the work force and technological upgradation go hand in hand. On the other hand, Mitra (2009) observed a decline in employment to value added ratio with a rise in manufacturing imports including technology.

With this background, the rest of the chapter is organized as follows. In the next section, we assess the impact of R&D and other firm specific factors on employment. Finally the findings are summarized in Sect. 3. This study uses the firm-level data in the manufacturing sector, compiled by ACEEQUITY for the period 1998 through 2010. ACEEQUITY is one of the agencies which collect from the firms' annual reports information on various aspects such as sales, assets, wages and salaries, exports and imports and expenditure on R&D. We have extracted information for 11 groups of industries. The number of firms in many of these industries is sizeable—in fact, most of the large firms are included in the study. However, the panel is not balanced as the information on all the variables for a given firm is not available for all the years.

7.2 Impact of R&D on Employment

Employment to sales ratio perceived as a rough proxy for labour requirement per unit of output has been regressed on R&D to sales ratio, exports to sales ratio, imports to sales ratio, assets to sales ratio and efficiency (or TFPG). In an alternative specification, employment to sales ratio has been replaced by log of employment, without changing the determinants.[33] This is pursued mainly to capture the view that labour per unit of real output (approximated by real sales) may not increase in response to R&D though the overall employment may.[34] The performance indicator is included to test if TFP growth, for example, results in higher output growth

---

[33]Wages are not considered because capital asset is included as a determinant. Since capital itself is a function of wage rental ratio multicollinearity and the problem of double counting would have emerged had we included wage rate in the function.

[34]If the rise in output is more than employment then labour per unit of output may decline in spite of an increase in overall employment.

relative to input growth including labour or alternately, does not affect employment though reduces the use of other inputs.

In the equation which includes technical efficiency as one of the determinants, the following three industry groups unravel a positive effect of R&D to sales ratio on employment: engineering (Industrial Equipment), household and personal products, pharmaceutical and drugs (Table 23). In the rest of the industries, R&D to sales ratio remains insignificant. Technical efficiency shows a negative effect on employment to sales ratio in the case of electronics component and a positive effect in engineering (industrial equipments) and remains insignificant in the rest of the industries.

The ratio of exports to sales is significant with a positive coefficient in three industries and negative only in one. Similarly, the imports to sales ratio show a significant value only in three industries and among them two are positive. Based on this it is difficult to generalize that trade contributes to employment generation. However, some of the labour-intensive sectors like consumer durables (domestic appliances) and household and personal products show a positive effect of both export to sales and import to sales on employment to sales. While higher exports lead to increased employment, imported inputs also tend to create employment, suggesting possibilities of complementary relationship between the imported inputs and skilled labour. Not any major improvement in results is obtained by redefining the dependent variable as log transformation of employment.[35]

As we replace technical efficiency by TFPG in the equation for employment to real sales ratio, the results relating to R&D/Sales ratio remain unchanged except for electronics component which now turns out to be negative and significant (Table 24). TFPG itself is significant only in two industries with a negative coefficient, implying higher growth in output relative to input growth. Electronics component and household and personal products show a positive effect of both exports and imports. Even after changing the dependent variable to log of employment both these industries continue to indicate the positive effect of trade. Also, after changing the dependent variable to log of employment, electronics component, engineering (industrial equipment) and leather show a positive effect of R&D to sales on employment with no negative effect in any of the other industries.[36]

On the whole, the R&D/sales ratio is not significant in a number of industries; however, the cases of positive impact are noteworthy.

As we drop TFPG or technical efficiency from the employment equation, the effect of R&D/sales on employment to sales turns out to be positive and significant in the following four industries: electric equipment, engineering (industrial equipments), household and personal products, pharmaceuticals and drugs. In the rest of the industries, the effect is statistically insignificant.

The export to sales ratio is positive in four industries and negative in two industries. Similarly, the import to sales is positive in three industries and negative

---

[35]Results not reported.

[36]Results not shown.

Table 23 Employment/sales and R&D/sales with TE-dependent variable: employment/sales

Industry	Model	R&D/sales	Export/sales	Import/sales	Asset/sales	TE	Constant	R^2/Adj. R^2	N
Consumer durables-domestic appliances	OLS	−60.906 (−0.33)	5.501* (1.96)	35.87** (4.03)	1.452 (0.40)	0.988 (0.15)	7.307 (1.61)	0.37	29
Consumer durables-electronics	OLS	309.098 (1.55)	−32.78** (−2.91)	3.074 (0.98)	−10.741 (−1.77)	21.641 (1.49)	2.868 (0.50)	0.70	18
Chemical	RE	30.674 (1.23)	−1.045 (−1.00)	−0.837 (−0.66)	1.856** (4.44)	1.002 (0.24)	3.063** (2.36)	0.07	186
Electric equipment	RE	203.766 (1.22)	−1.135 (−0.22)	−0.448 (−0.05)	1.484 (0.81)	−31.656 (−1.14)	9.350** (2.86)	0.03	96
Electronics component	OLS	31.550 (0.18)	32.149** (2.94)	19.948 (1.66)	0.172 (0.64)	−83.524** (−3.53)	29.970** (4.19)	0.41	32
Engineering	RE	161.433 (0.49)	−1.057 (−0.19)	−2.679 (−0.42)	8.776* *(3.37)	−21.047 (−1.22)	10.140* *(2.68)	0.40	45
Engineering construction	OLS	3968.547 (1.21)	−4299.38 (−1.45)	17.053 (0.82)	0.687 (0.10)	−40.983 (−1.12)	40.997 (1.20)	0.36	7
Engineering—industrial equipments	OLS	1431.9** (2.62)	2.444 (0.31)	−91.238** (−3.37)	8.080** (3.57)	104.567** (2.88)	−28.425** (−2.10)	0.68	31
Household and personal products	RE	374.736** (3.00)	18.092** (2.13)	41.933** (2.71)	2.105 (1.10)	−27.171 (−1.31)	14.116** (2.31)	0.02	46
Leather	RE	1799.993 (0.92)	15.840 (1.62)	−30.546 (−0.93)	−5.998 (−0.33)	−29.798 (−1.31)	27.965 (1.63)	0.31	26
Pharmaceuticals and drugs	FE	56.84** (5.53)	−3.520 (−1.37)	1.738 (0.42)	8.70** (318.97)	−41.86 (−1.45)	11.692** (3.33)	0.97	499

Note Figure in parenthesis are t-values for FE model and OLS and z-value for RE model. ** and * denote 5 and 10 % level of significance, respectively. *FE* denotes fixed effect model: *RE* denotes random effect model: *OLS* denotes ordinary least square. Adj. R^2 is calculated only for OLS

Table 24 Employment/sales and R&D/sales with TFPG-dependent variable: employment/sales

Industry	Model	R&D/Sales	Export/Sales	Import/Sales	Asset/Sales	TFPG	Constant	R^2/Adj. R^2	N
Consumer durables-domestic appliances	OLS	−49.604 (−0.26)	5.494 (1.68)	35.483** (4.04)	1.390 (0.38)	1.541 (0.05)	7.933** (4.69)	0.37	29
Consumer durables-electronics	OLS	119.961 (0.50)	−32.922* (−2.72)	−1.491 (−0.69)	−3.828 (−0.42)	−1.265 (−1.10)	10.801** (7.99)	0.67	18
Chemical	FE	25.810 (0.98)	−0.244 (−0.20)	−0.115 (−0.09)	1.964** (4.43)	−1.069 (−0.24)	3.216** (6.56)	0.03	186
Electric equipment	RE	171.269 (1.03)	0.343 (0.07)	−3.024 (−0.37)	2.482 (1.45)	4.331 (0.37)	6.066** (2.76)	0.005	96
Electronics component	RE	−425.871* (−1.92)	14.137* (1.94)	22.107** (2.24)	1.062** (3.22)	−49.693** (−2.88)	5.626** (2.79)	0.47	30
Engineering	RE	127.925 (0.38)	−1.050 (−0.19)	−5.082 (−0.78)	9.592** (3.92)	−14.093 (−1.08)	7.162** (2.73)	0.28	45
Engineering construction	OLS	−1633.149 (−0.24)	1167.04 (0.27)	32.055 (1.49)	7.146 (1.05)	28.362 (0.69)	1.882 (0.64)	0.02	7
Engineering—industrial equipments	OLS	2076.12** (3.56)	10.245 (1.16)	−76.600** (−2.45)	4.745* (1.74)	−5.985 (−0.08)	8.787* (1.79)	0.58	31
Household and personal products	RE	403.708** (2.88)	19.15** (2.07)	38.48** (2.23)	2.055 (1.01)	−9.267 (−0.81)	6.910** (3.48)	0.02	46
Leather	OLS	1065.961 (0.65)	22.45** (2.64)	−28.044 (−1.04)	9.830 (0.62)	−128.113** (−3.22)	5.005 (0.44)	0.38	26
Pharmaceuticals and drugs	RE	58.454** (5.93)	−5.212** (−2.29)	0.099 (0.03)	8.701** (323.13)	0.970 (0.10)	7.278** (5.09)	0.98	499

Note Figure in parenthesis are t-values for FE model and OLS and z-value for RE model. ** and * denote 5 % and 10 % level of significance, respectively. *FE* denotes fixed effect model: *RE* denotes random effect model: *OLS* denotes ordinary least square. Adj. R^2 is calculated only for OLS

Table 25 Employment/sales and R&D/sales without performance indicator dependent variable—employment/sales

Industry	Model	R&D/sales	Export/sales	Import/sales	Asset/sales	Constant	R^2/Adj. R^2	N
Consumer durables-domestic appliances	OLS	-42.461 (-0.30)	5.834** (2.58)	36.029** (4.77)	0.684 (0.25)	8.011** (5.44)	0.41	30
Consumer durables-electronics	OLS	254.966 (1.24)	-38.581** (-3.49)	-0.612 (-0.30)	-11.041 (-1.74)	11.256** (8.67)	0.67	18
Chemical	FE	25.097 (0.96)	-0.267 (-0.22)	-0.085 (-0.07)	1.961** (4.44)	3.224** (6.61)	0.03	186
Electric equipment	RE	160.121 (0.98)	0.818 (0.17)	-4.441 (-0.59)	2.532 (1.57)	6.356** (3.17)	0.004	97
Electronics component	RE	390.092** (2.58)	39.079** (2.43)	13.932 (1.13)	0.311 (1.08)	4.041 (0.86)	0.17	32
Engineering	RE	167.599 (0.52)	-1.038 (-0.19)	-3.338 (-0.54)	10.296** (4.40)	6.249** (2.55)	0.28	47
Engineering construction	OLS	252.464 (1.28)	9.079* (2.49)	28.312** (2.92)	5.148 (1.58)	3.081** (3.20)	0.46	10
Engineering—industrial equipments	OLS	2068.919** (3.67)	10.132 (1.19)	-76.14** (-2.53)	4.861** (2.18)	8.708* (1.85)	0.60	31
Household and personal products	RE	355.182** (2.79)	15.870* (1.84)	44.558** (2.82)	2.594 (1.38)	6.597** (3.81)	0.03	47
Leather	FE	221.470 (0.25)	-0.415 (-0.05)	9.806 (0.54)	41.032** (4.75)	3.798 (0.46)	0.01	26
Pharmaceuticals and drugs	RE	124.543** (113.87)	-10.507** (-4.02)	-6.449 (-1.50)	8.706** (266.35)	8.473** (5.75)	0.98	507

Note Figure in parenthesis are *t*-values for FE model and OLS and *z*-value for RE model. ** and * denote 5 % and 10 % level of significance, respectively. *FE* denotes fixed effect model: *RE* denotes random effect model: *OLS* denotes ordinary least square. Adj. R^2 is calculated only for OLS equation

in only one. Interestingly, all these three industries showing positive effect of imports, also show the positive effect of exports (e.g. Consumer durables-domestic appliances, Engineering construction, Household and personal products). In five industries, the asset to sales ratio shows a positive effect on employment (Table 25).

In several studies, employment is taken to be a function of value added and wage rate to estimate the growth and wage elasticity of employment. Following the same logic, we may regress log of employment on log of real sales, real wage rate (derived by deflating the nominal figures by the consumer price index for industrial workers), and in addition real RND (deflated by the price index for machinery). Since R&D/Sales ratio has a highly limited variation across companies and over time, log of R&D may be considered to be more suitable.

In this specification (Table 26), log R&D turns out to be significant with a positive effect in a number of industries (seven) and the elasticity of employment with respect to R&D is seen to be highest in consumer durables (around 0.3). In two

Table 26 Partial elasticity of employment with respect to sales, wages and R&D dependent variable—LnEmployment

Industry	Model	LnSales	LnR&D	LnWage rate	Constant	R^2/Adj. R^2	N
Consumer durables-domestic appliances	RE	0.784** (15.03)	0.051** (2.02)	−0.470 (−1.40)	−2.36** (−2.78)	0.89	78
Consumer durables-electronics	OLS	0.498** (8.74)	0.297** (8.48)	−0.325 (−0.41)	−0.448 (−0.35)	0.95	33
Chemical	FE	0.497** (14.99)	0.015 (0.74)	−0.745** (−8.59)	1.180** (4.13)	0.71	586
Electric equipment	FE	0.484** (10.34)	0.070** (3.98)	−0.821** (−4.61)	1.55** (4.49)	0.85	225
Electronics component	RE	0.581** (14.03)	0.066** (2.38)	−1.348** (−6.79)	1.358** (2.69)	0.92	101
Engineering	FE	0.477** (11.79)	0.023 (0.96)	−1.051** (−16.56)	2.712** (7.37)	0.77	186
Engineering construction	RE	0.825** (10.30)	0.007 (0.28)	−1.012** (−2.34)	−1.760** (−2.29)	0.73	83
Engineering—industrial equipments	RE	0.732** (17.16)	0.017 (0.67)	−0.861** (−8.28)	−0.559 (−1.40)	0.84	98
Household and personal products	RE	0.745** (12.07)	0.065** (2.62)	−0.904** (−5.30)	−0.476 (−0.70)	0.95	61
Leather	RE	0.802** (6.96)	0.153** (2.19)	0.124 (0.17)	−3.142** (−2.45)	0.67	54
Pharmaceuticals and drugs	FE	0.443** (18.30)	0.117** (7.67)	0.007 (0.07)	0.811** (3.55)	0.82	1194

Note Figure in parenthesis are *t*-values for FE model and OLS and *z*-value for RE model. ** and * denote 5 % and 10 % level of significance, respectively. *FE* denotes fixed effect model: *RE* denotes random effect model: *OLS* denotes ordinary least square. Adj. R^2 is calculated only for OLS

other industries (Leather and Pharmaceutical), it is again a little above 0.1. In electric equipment, electronics component and household and personal products also the estimate is closer to 0.1.

7.3   Conclusion

This study based on the firm-level data for 11 groups of industries examined the impact of R&D on employment, taking R&D expenditure as a broad proxy for innovation pursuits of the firms.

As far as the impact of R&D as a percentage of sales on employment is concerned, the study noted a positive relationship only in a few industries. This has been tested with and without controlling for the performance indicator, which does not show any strong effect on employment. Since R&D to sales ratio is seen to have a very limited variation, a different specification has been used which captures the partial elasticity of employment with respect to growth, wage rate and R&D expenditure. A number of industries reported a positive effect of R&D on employment in this specification. Also, some of the labour-intensive industries revealed a higher elasticity of employment with respect to R&D expenditure. Combining the results of all the specifications, we may conclude that this is indicative of a positive effect of R&D on employment in absolute sense though the employment content per unit of output does not increase in many industries. Recalling that R&D expenditure in the Indian context is sloppy and does not necessarily mean actual innovation, the positive effect of R&D on employment in absolute sense is noteworthy. Processing of byproducts and efforts pursued to bring in an improvement in product quality and efficiency are some of the striking features of R&D expenditure, which may be resulting in employment gains. Even when capital-intensive technology is adopted by the firms, R&D expenditure has the potentiality to generate employment as it means additional activities without involving additional capital. From this point of view, the study has important policy implications in terms of new schemes which can provide incentives to encourage R&D expenditure. Effective supervision can also channelize R&D expenditure towards actual innovation of technology that is appropriate for labour surplus countries. On the whole, the concern that innovation causes employment loss is rather exaggerated.

8   Conclusion

The performance of the organized manufacturing in India based on gross value added growth rate showed marked improvement in the nineties compared to the earlier period. In terms of growth rates both workers and total persons employed increased from a mere 1 % per annum during the deregulated regime (1984–1985 to 1990–1991) to around 3 % per annum over the 1990s, though this growth rate was marginally above the growth rate that was experienced during the regulated regime (1973–1974 through 1984–1985). The increase in the employment growth rate in the organized manufacturing in the nineties, particularly between 1990–1991 and 1995–1996, could also be explained by the huge expansion that took place in the early reform period.

Subsequently, the gross value added growth rate continued to be a little above 9 % per annum over 1998–1999 through 2007–2008. However, the employment growth rate during this phase decelerated marginally in comparison to the nineties (prior to 1998–1999) experience, and more so in the case of employees other than workers. Labour productivity defined as the value added per person employed grew at almost 7 % per annum. Wages per worker remained almost stagnant while the remuneration per person shot up significantly, implying a substantial growth in the salaries per employee (excluding workers). The growth rate in fixed capital decelerated to almost half, resulting in a sluggish growth in capital-labour ratio. Since Indian entrepreneurs had already accumulated a huge stock of capital—evident from a sustained growth in fixed capital over the preceding period—the decline in the growth rate during 1997–1998 to 2007–2008 does not come as a surprise.

Industries which dominated in terms of employment share did not necessarily unravel a fast employment growth. Over this phase, several industries, however, witnessed rapid value added growth. Industries which did not perform well in terms of value added did not perform well in terms of employment either. There is a strong positive correlation between the average value added growth and total employment growth measured across all the three-digit manufacturing groups, implying growth is essential for employment generation. However, not necessarily rapid value added growth has resulted in faster employment growth. Total employment increased only at around 2.6 % per annum at the aggregate level, over the period 1998–1999 through 2007–2008.

Decomposing total employment in terms of workers and employees (other than workers) several industries are seen to have experienced a negative growth rate in the latter category, though value added growth in these industries was more than 5 % per annum. Four clusters are discernible: (a) low employment and low value added growth, (b) low employment but high value added growth, (c) moderate employment and high value added growth and (d) high employment and high value added growth. Decomposition of value added in terms of employment and productivity shows that only a few groups experienced a rapid productivity growth of at least 5 % per annum and employment growth of at least 4 % per annum, simultaneously. Capital-intensive technology leading to rapid labour productivity growth naturally reduces the extent of employment generation.

For the aggregate organized manufacturing sector the elasticity of total employment with respect to value added is 0.43 and the elasticity of workers with respect to value added is 0.35. Industries which recorded an employment elasticity of up to 0.4 are many, though only a handful of them had an employment elasticity of more than 0.55. These industries may be targeted for providing a boost to employment growth but given their share it is unlikely that they will succeed in raising the overall employment growth of the manufacturing sector in a significant manner as specified in the NMP.

The impact of remuneration per employee (or wages per worker) on total employment (or workers) is not statistically significant. The correlation between the average growth in workers' productivity and wages is again negligible. Since wages do not constitute a large component of the total cost, a more careful analysis has to be pursued to identify what restricts industrial expansion and employment creation rather than simply blaming the strict labour laws. Wage share can be interpreted as the wage rate to productivity ratio, implying that if the numerator is greater than the denominator it reflects gross inefficiency. However, not a single case is seen where the ratio is greater than unity. Thus, the popular views on unionization and inefficiency are apparently exaggerated.

Even within the ASI sector, the proportion of contractual workers to total workers (directly employed plus those appointed through the contractors) has increased steadily over time. Second, in many industries the proportion has been extremely high. The third observable pattern relates to the ratio of female to male workers in the category of directly recruited workers, which is significant in a group of industries. Since female wages are usually believed to be lower than that of the males, feminization of the work force may be treated as an indicator of labour market flexibility that the employers have chosen as a means to reduce labour cost. On the whole, mechanisms to initiate flexibility are already in motion even without labour market deregulation being carried out formally.

In the context of the unorganized manufacturing, the growth rate in value added has been much faster than the growth rate in number of workers and number of enterprises during 1989–1990 to 2005–2006. Over the more recent period, i.e. 2005–2006 through 2010–2011, employment growth has been mostly negative in the own account enterprises. However, in the establishments it was a little below 2 % per annum though the aggregate employment figure for all establishments and own account enterprises turns out to be negative.

Wages or emoluments per worker in the unorganized manufacturing grew at a rapid pace. Even across various groups of industries the elasticity of wages with respect to productivity turns out to be unity or more than unity, implying wage growth has been more than the productivity growth. However, we need to recognize the fact that the base year wage in the unorganized sector has been much lower than that in the organized sector and, therefore, a higher growth rate corresponding to the unorganized manufacturing sector does not come as a surprise. Notwithstanding these explanations, we may note that globalization has resulted in a higher degree of mobility across activities and regions and in an attempt to prevent high labour turnover cost it is natural for the employers to offer higher wages. Hence, part of the rapid growth in wages in the unorganized sector is in response to the changing economic environment. Also, subcontracting from large units is believed to have raised the supply of work consignments in the unorganized sector which possibly contributed to rapid wage growth. Subcontracting from the large units requires quality control which can be assured partly through improving the tenure of the workers, i.e. by reducing the labour turnover rate and introducing regularity in the informal/unorganized sector, which is possible only by raising the wages.

As regards inter-industry linkages, we may conclude that some of the labour-intensive industries do not have strong inter-industry linkages whereas some others like textile etc. do have inter-connections, implying that industrial deceleration in the heavy goods sector can reduce the input demanded from the labour as well as capital-intensive sub-sectors. Thus, the growth and employment in the labour-intensive sub-sectors may suffer which in turn may affect adversely the overall employment growth in the manufacturing sector. This may also have spill-over (negative) effects on the rest of the economy and the pace of employment generation and the effectiveness of the industrial sector in reducing poverty.

The index representing the difference between the skill level of the population and the workers is estimated at 73.11. In other words, the difference between the skill levels of the potential labour supply and those already working is sizeable. The inter-temporal skill mismatch index as estimated from the distribution of workers in each of the activities across various skill levels is again substantial. Relatively speaking, the skill gap in the manufacturing sector is seen on the high side compared to some of the tertiary activities, indicating that over time jobs in the manufacturing are becoming more skill-biased. In the earnings function, the dummies representing different educational levels are significant, suggesting positive returns to education. Thus, those who are able to acquire skill are able to secure employment with higher levels of earnings as demand exists for the skilled variety of labour due to rise in skill intensity of the technology. Absorption of the unskilled labour is, therefore, a major challenge.

The important policy implications are as follows:

Industries which have experienced rapid growth in both value added and employment need to be given priority. Those with high employment elasticity with respect to growth can be encouraged for enhancing manufacturing employment. The industries which are highly labour intensive and have recorded a sizeable employment growth naturally deserve a greater attention.[37] Even the industries which are labour intensive but are not able to grow at a rapid pace fall into the domain of a closer scrutiny.

Even in the capital-intensive sector possibilities have to be explored if some of the phases in the production process can be carried out on the basis of labour intensive methods. Since labour productivity growth and not employment has been the major contributing factor to value added growth, emphasis has to be laid on reducing the capital accumulation process if manufacturing has to be the engine of employment generation.

Labour market deregulation may not bring in rapid employment growth because the responsiveness of employment with respect to wages is not statistically significant across a number of industry groups. The contractualization is already on the

---

[37]181, 182, 192, 191, 369, 281, 372, 371, 172, 361, 273, 289, 312, 173, 319, 343.

rise in several industries even without much deregulation measures carried out in black and white. In spite of that if employment could not pick up sizably, other constructive routes to employment generation need to be pursued instead of blaming the labour market laws univocally.

Improvement in employability is an important consideration from policy point of view. For this, skill formation is an essential prerequisite which can be attained by accessing quality education and participating in institutions which impart training in skill formation. Such technical institutions, particularly which provide diplomas, are however few in number and thus government initiative is indeed crucial. From the point of view of the quality of vocational education, again greater efforts are called for. Besides, on the job training is another important way of eliminating skill mismatches.

Realizing the importance that over the next decade, India has to create gainful employment opportunities for a large section of its population, with varying degrees of skills and qualifications, the manufacturing sector is expected to be the engine of this employment creation initiative. Apart from the employment imperative, the development of the manufacturing sector is critical from the point of view of ensuring a sustainable economic growth in India. Thus, with the objective of developing Indian manufacturing sector to reflect its true potential, the Department of Industrial Policy and Promotion (DIPP), Ministry of Commerce and Industry, has embarked on creating a policy environment that would be suitable for the manufacturing sector to grow rapidly. Keeping in view, the importance of the employment-industrialization-policies as mentioned above and also the fact that India has not been able to generate employment opportunities in the organized/formal manufacturing sector on a large scale, the NMP comes as a silver lining.

In the backdrop of a global recession and large job losses if corrective steps are not taken India's situation can be worse off. From this perspective, the recently cleared NMP promises to create a 100 million more jobs and contribute 25 % to country's GDP in a decade. In the face of dampening demand and rising cost of capital, the experts in the policy circle believed that it can change the fate of manufacturing in India and turnaround the overall economy. The policy addresses in great detail the environment and regulatory issues, labour laws and taxation, but it is the proposed creation of National Manufacturing Investment Zones (NIMZs) or clustering of manufacturing units that is treated as a unique way of integrating the industrial infrastructure and achieve economies of scale. NIMZs will be developed as integrated industrial townships with world class infrastructure and land use on the basis of zoning, clean and energy efficient technology with a size of at least 5000 ha. The NIMZs will be on the non-agricultural land with adequate water supply and the ownership will be with the state government. It aims at introducing flexibility in the labour market by offering greater freedom to the employers while hiring and firing. It also enables the sunset industrial units to follow a simplified exit mechanism. At the same time, it insists on workers' rights which run the risk of being compromised in the name of flexibility.

An important feature of the manufacturing policy is its financial and development incentives to the small and medium enterprises (SMEs). On the whole, the policy, promises to increase the share of manufacturing sector to the country's gross domestic product to 25 % from existing 16. However, the national manufacturing policy's objective of raising the industrial employment to an unprecedented level may not be realized as the organized manufacturing employment comprises only a fraction of the total manufacturing employment.

It may be therefore useful to consider the employment potential of the unorganized manufacturing sector as well and tap the potentials to create quality-employment in this sector. SMEs need to undergo an innovative revolution in terms of scale of operations, technology, financing and ways to upgrade skills of workers. Since labour-intensive sectors like food processing, apparels and textiles, leather and footwear contribute to over 60 % of SMEs' employment (Kant 2013), greater focus on the labour-intensive sectors will enable productive absorption of surplus unskilled labour. Though our study did not deal with the regional profile of the labour market and aspects relating to inter-spatial industrial growth disparity, the policy initiatives need to give top priority to labour-intensive goods based industrial growth in regions characterized by greater magnitudes of unskilled labour and insignificant industrialization.

Issues relating to infrastructure shortage, constraints on energy supply, sluggish exports growth and poor performance of labour-intensive exportable goods sector, the lack of innovations required for developing appropriate technology and bureaucratic and administrative rigidities in areas where they tend to hamper growth and employment or attract foreign investment are undoubtedly important though an empirical investigation of all of that remained outside the ambit of the present study.

As far as the impact of R&D as a percentage of sales on employment is concerned, the study noted a positive relationship only in a few industries. This has been tested with and without controlling for the performance indicator, which does not show any strong effect on employment. Since R&D to sales ratio is seen to have a very limited variation, a different specification has been used which captures the partial elasticity of employment with respect to growth, wage rate and R&D expenditure. A number of industries reported a positive effect of R&D on employment in this specification. Also, some of the labour-intensive industries revealed a higher elasticity of employment with respect to R&D expenditure. Combining the results of all the specifications, we may conclude that this is indicative of a positive effect of R&D on employment in absolute sense though the employment content per unit of output does not increase in many industries. The study has important policy implications in terms of new schemes which can provide incentives to encourage R&D expenditure. Effective supervision can also channelize R&D expenditure towards actual innovation of technology that is appropriate for labour surplus countries. With a focus on pro-poor growth issues if innovation is undertaken domestically, employment growth is certain to take place along with output growth.

References

Acemoglu D, Zilibotti F (2001) Productivity differences. Quart J Econ 116:563–606

Ahsan A, Pagés C (2009) Are all labor regulations equal? Evidence from Indian manufacturing. J Comp Econ 37(1):62–75

Azeez AE (2006) Domestic capacity utilization and import of capital goods: substitutes or complementary? Evidence from Indian capital goods sector. In: Tendulkar SD et al (eds.) India: Industrialisation in a reforming economy (Essays for K.L.Krishna). Academic Foundation, New Delhi, pp 141–156

Banga R, Goldar B (2004) Contribution of services to output growth and productivity in Indian manufacturing: pre and post reforms. Econ Polit Wkly 42:2769–2777

Berg J, Cazes S (2007) The doing business indicators: Measurement issues and political implications (Economic and Labour Market Papers 6). Employment analysis and research unit, economic and labour market analysis department, International Labour Office, Geneva

Berman E, Machin S (2000) Skill-biased technology transfer: evidence of factor biased technological change in developing countries, Mimeo. Available at http://dss.ucsd.edu/~elib/glob.pdf

Berman E, Machin S (2004) Globalization, skill-biased technological change and labour demand. In: Lee E, Vivarelli M (eds) Understanding globalization, employment and poverty reduction. Palgrave Macmillan, New York, pp 39–66

Besley T, Burgess R (2004) Can labor regulation hinder economic performance? Evidence from India. Q J Econ 119(1):91–134

Bhalotra S (1998) The puzzle of jobless growth in Indian manufacturing. Oxford Bull Econ Stat 60 (1):5–32

Bhattacharjea A (2006) Labour market regulation and industrial performance in India: a critical review of the empirical evidence. Indian J Labour Econ 49(2):211–232

Bishop D (2012) Firm size and skill formation processes: an emerging debate. J Educ Work 25 (5):507–521

Bogliacino F (2014) Innovation and employment: a firm level analysis with European R&D scoreboard data. Economia 15:141–154

Bogliacino F, Vivarelli M (2012) The job creation effect of R&D expenditures. Aust Econ Pap 51:96–113

Bogliacino F, Piva M, Vivarelli M (2012) R&D and employment: an application of the LSDVC estimator using European micro-data. Econ Lett 116:56–59

Brady D, Kaya Y, Gereffi G (2011) Stagnating industrial employment in Latin America. Work Occup 38(2):179–220

Braun S (2008) Economic integration, process and product innovation, and relative skill demand. Rev Int Econ 16:864–873

Chakravarty S (1987) Development planning: the Indian experience. Oxford University Press, New Delhi

Castellani D, Zanfei A (2006) Multinational firms, innovation and productivity. Edward Elgar Cheltenham, UK

Chandrasekhar CP (1992) Investment behaviour, economies of scale and efficiency in an import substituting regime: a study of two industries. In: Ghosh A et al (ed) Indian industrialisation, structure and policy issues. Oxford University Press, New Delhi

Choi J-Y, Yu ESH, Jin JC (2002) Technical progress, urban unemployment, outputs, and welfare under variable returns to scale. Int Rev Econ Finance 11:411–425

Comyn P (2012) Skill intensity and skills development in bangladesh manufacturing enterprises. J Educ Work 1–29

Crespi F, Pianta M (2006) Demand and innovation in productivity growth. http://www.ec.unipg.it/DEFS/uploads/crespiantaproductivity.pdf

Csacuberta C, Fachola G, Gandelman N (2004) The impact of trade liberalisation on employment, capital and productivity dynamics, evidence from the uruguayan manufacturing sector

Das DK, Kalita G (2009a) Do labor intensive industries generate employment? Evidence from firm level survey in India (Working Paper No. 237). Indian Council for Research on International Economic Relations, New Delhi

Das DK, Kalita G (2009b) Are labour intensive industries generating employment in India? Evidence from the firm level survey. Indian J Labour Econ 52(3)

Dasgupta S, Singh A (2005) Will services be the new engine of economic growth in India? Working Paper No. 310, Centre for Business Research, University of Cambridge, Cambridge

Datta RC (2003) Labour market—social institution, economic reforms and social cost. In: Uchikawa S (ed) Labour market and institution in India. Manohar Publications, New Delhi

Dellas H, Koubi V (2001) Industrial employment, investment equipment, and economic growth. Econ Dev Cult Change. 49(4):867–81 (University of Chicago Press)

Desjardins R, Rubenson K (2011) An analysis of skill mismatch using direct measures of skills. OECD Education Working Papers No. 63

D'souza E (2008) Labour market institutions in India: their impact on growth and employment (ILO Asia-Pacific Working Paper Series). International Labour Organisation, Sub-regional Office for South Asia, New Delhi

Edwards AC, Edwards S (1994) Labour market distortions and structural adjustment in developing countries. In: Horton S, Kanbur R, Mazumdar D (eds) Labour market in an era of adjustment EDI development studies, vol 1. World Bank, Washington, DC, pp 105–145

Estevão M, Tsounta E (2011) Has the great recession raised U.S. structural unemployment? IMF Working Paper, WP/11/105. International Monetary Fund

Evenson R, Westphal LE (1995) Technological change and technology strategy ch.37. In: Behrman J, Srinivasan TN (eds) Handbook of development economics, vol 3A. Amsterdam, North-Holland, (Chapter 7), pp 2209–2229

Fallon PR, Lucas REB (1991) The impact of changes in job security regulations in India and Zimbabwe. World Econ Rev 5(3):395–413

Froumin I, Divakaran S, Tan H, Savchenko Y (2007) Strengthening skills and education for innovation. In: Dutz MA (ed) Unleashing India's innovation: toward sustainable and inclusive growth. World Bank, pp 129–146

Ghose AK (1994) Employment in organised manufacturing in India. Indian J Labour Econ 37 (2):141–162

Goldar B (2000) Employment growth in organised manufacturing in India. Econ Polit Wkly 35:1191–1195

Goldar B (2011a) Growth in organised manufacturing employment in recent years. Econ Polit Wkly, XLVI(7):20–23

Goldar B (2011b) Productivity in Indian manufacturing in the post-reform period. Institute of Economic Growth, New Delhi

Goldar B, Agarwal SC (2005) Trade liberalisation and price cost margin in Indian industries. Dev Econ XLIII-3:346–373

Goldar B, Mitra A, Kumari Anita (2011) Performance of unorganised manufacturing in the post-reforms period. In: Das K (ed) Micro and small enterprises in India, the era of reforms. Routledge, New Delhi

Guangrong T, Yuanyuan L (2009) The affection of independent innovation on employment. Manag Sci Eng 3:36–40

Hajela R (2012) Shortage of skilled workers: A paradox of the Indian Economy, SKOPE, Research Paper No. 111, ESRC Centre on skills, knowledge and organisational performance, COMPAS University of Oxford

Hall BH, Lotti F, Mairesse J (2008) Employment, innovation and productivity: evidence from Italian microdata. Ind Corp Change 17:813–839

Harrison RJ, Jaumandreu J, Mairesse J, Peters B (2014) Does innovation stimulate employment? A firm level analysis using comparable micro-data from four European countries. Int J Ind Organ 35:29–43

Hasan R (2002) The impact of imported and domestic technologies on the productivity of firms: panel data evidence from Indian manufacturing firms. J Dev Econ 69:23–49

Hasan R (2003) The impact of trade and labour market regulations on employment and wages: evidence from developing countries. In: Hasan R, Mitra D (eds) The impact of trade on labour: issues, perspectives and experiences from developing Asia. North Holland, Elsevier, Amsterdam

Hasan R, Mitra D, Ramaswamy KV (2003) Trade reforms, labor regulations and labor demand elasticities: empirical evidence from India (NBER Working Paper, No. w9879). National Bureau of Economic Research, Inc, Cambridge

Hijzen A, Swaim P (2007) Does offshoring reduce industry employment? Research paper series globalisation and labour markets research paper 2007/24. Leverhulme Centre, The University of Nottingham

ILO-SAAT (1996) Economic reforms and labour policies in India. ILO-SAAT, New Delhi

James J (1993) New technologies, employment and labour markets in developing countries. Dev Change 24:405

Johanson RK (2004) Implications of globalization and economic restructuring for skills development in Sub-Saharan Africa (Working Paper No. 29). ILO, Geneva

Kaldor N (1967) Strategic factors in economic development. Cornell University Press, Ithaca

Kannan KP, Raveendran G (2009) Growth sans employment: a quarter century of Jobless Growth in India's organised manufacturing. Econ Polit Wkly XLIV(10)

Kant A (2013) For a manufacturing revolution. The Times of India

Kato A, Mitra A (2008) Imported technology and employment: evidence from panel data on Indian manufacturing firms. In: Hashim SR, Siddharthan NS (eds) High-tech industries, employment and global competitiveness. Routledge, New Delhi, pp 180–194

Kelley AC et al (1972) Biased technological progress and labour force growth in a dualistic economy. Q J Econ 86:426–447

Kuznets S (1966) Modern economic growth: rate, structure and spread. Yale University Press

Lee E, Vivarelli M (2006) The social impact of globalisation in the developing countries. CSGR Working Paper No 199/06

Majumdar D, Sarkar S (2004) Reforms and employment elasticity in organised manufacturing. Econ Polit Wkly XXXIX(27)

Mayer J (2000) Globalization, technology transfer and skill accumulation in low-income countries (WIDER Project, No. 150). UNCTAD, Geneva. Available at http://www.unctad.org/en/docs//dp_150.en.pdf

Mehrotra S, Ankita G, Sahoo BK (2013) Estimating India's skill gap on a realistic basis for 2022. Econ Polit Wkly XLVIII(13):102–111

Mitra A (2009) Technology import and industrial employment: evidence from developing countries. Labour 23(4):697–718

Mitra A (2013) Insights into inclusive growth, employment and wellbeing in India. Springer, Berlin

Mitra A, Bhanumurthy NR (2006) Economic growth, employment and poverty: a study of manufacturing, construction and tertiary sectors in India. Employment strategy department, international labour organisation, Geneva

Mitra A, Schmid J (2008) Growth and poverty in India: emerging dimensions of the tertiary sector. Serv Ind J 28(8):1–22

Mitra A, Varoudakis A, Veganzones-Varoudakis MA (2002) Productivity and technical efficiency in Indian states manufacturing: the role of infrastructure. Econ Dev Cult Change 50(2):395–426

Mitra A Jha A (2015) Innovation and employment: a firm level study of Indianindustries. Eurasian Bus Rev 5(1):45–71

Mureithi L (1974) Demographic and technological variables in Kenya's employment scene. Eastern Africa Econ Rev 6:27–43

Nagaraj R (1994) Employment and wages in manufacturing industries: trends, hypotheses and evidence. Econ Polit Wkly 29(4):177–186

Nagaraj R (2000) Organised manufacturing employment. Econo Polit Wkly XXXV(38)

Nagaraj R (2004) Fall in organised manufacturing employment: a brief note. Econ Polit Wkly 3:3387–3390

Nagaraj R (2007) Labour market in India. Paper Presented in seminar on Labour Markets in Brazil, China and India

Nagaraj R (2011) Growth in organised manufacturing employment in recent years. Econ Polit Wkly 46:20–23

Pack H, Todaro M (1969) Technological transfer, labour absorption, and economic development. Oxford Econ Pap 21:395–403

Pandit BL, Siddharthan NS (2006) MNEs, product differentiation, skills and employment: lessons from the Indian experience. In: Hashim SR, Siddharthan NS (eds) High-tech industries, employment and global competitiveness. Routledge, New Delhi, pp 165–179

Papola TS (1994) Structural adjustment, labour market flexibility and employment. Indian J Labour Econ 37(1):3–16

Pianta M (2005) Innovation and employment. In: Fagerberg J, Mowery D, Nelson RR (eds) Handbook of innovation. Oxford University Press, Oxford, pp 568–598

Piva M (2003) The impact of technology transfer on employment and income distribution in developing countries: a survey of theoretical models and empirical studies. International labour office, policy integration department, international policy group, Working Paper n.15, ILO, Geneva

Pradhan BK, Saluja MR, Sharma AK (2013) Social accounting matrix for India 2007–08. IEG Working Paper

Ramaswamy KV (1994) Small-scale manufacturing industries: some aspects of size, growth and structure. Econ Polit Wkly 29(9):M13–M23

Richardson S (2007) What is a skill shortage?, National Institute of Labour Studies Flinders University, Australian Government

Rodrik D (1997) Has globalization gone too far?. Institute for International Economics, Washington, DC

Saviotti PP, Pyka Andreas (2004) Economic development, variety and employment. Rev Économique 55:1023–1049

Schumpeter JA (1939) Business cycles: a theoretical, historical and statistical analysis of the capitalist process. McGraw-Hill, New York

Schumpeter JA (1961) The theory of economic development. Oxford University Press, New York

Szirmai A, Verspagen B (2011) Manufacturing and economic growth in developing countries. United Nations University, UNU-MERIT, Working Paper Series, 2011–069

Tilak JBG (2003) Higher education and development in Asia. J Educ Plann Adm XVII(2):151–173

Thomas JJ (2002) A review of Indian manufacturing. In: Parikh K, Radhakrishna R (eds) India Development Report 2002. Oxford University Press, New Delhi

Sankaran U, Abraham V, Joseph KJ (Undated) Impact of trade liberalisation on employment: the experience of India's manufacturing industries

UNIDO (2005) Productivity in developing countries: trends and policies. United Nations Industrial Development Organisation, Vienna

Vivarelli M (1995) The economics of technology and employment: theory and empirical evidence. Elgar, Cheltenham

Vivarelli M (2011) Innovation, employment and skills in advanced and developing countries: a survey of the literature. IDB Publications 61058, Inter-American Development Bank

Vivarelli M (2013) Technology, employment and skills: an interpretative framework. Eurasian Bus Rev 3:66–89

Wood A (1997) Openness and wage inequality in developing countries: The Latin American challenge to East Asian conventional wisdom. World Bank Econ Rev 33:33–57

World Bank (1989) India: poverty, employment and social services: a world bank country study. World Bank, Washington DC

Appendix

See Tables A.1, A.2, A.3, A.4, A.5, A.6, A.7, A.8, A.9, A.10 and A.11.

Table A.1 ASI code, name of the industry groups, WPI used to deflate the value added, distribution of value added and employment

ASI code	Industry name	WPI used (1993–94 Base) to deflate value added	Average share in total value added (%)	Average share in total employment (%)
151	Production, processing and preservation of meat, fish, fruit vegetables, oils and fats	b. Canning, preserving and processing of fish	1.13	1.92
152	Manufacture of dairy product (production of raw milk is classified in class 0121)	a. Dairy products	0.77	0.98
153	Manufacture of grain mill products, starches and starch products, and prepared animal feeds	c. Grain mill products	1.20	3.75
154	Manufacture of other food products	k. Other food products n.e.c	3.18	8.16
155	Manufacture of beverages	a. Wine industries	1.03	1.04
160	Manufacture of tobacco products (tobacco related products are also included while preliminary processing of tobacco leaves is classified in class 0111)	d. Manufacture of Bidi, Cigarettes, Tobacco and Zarda	0.01	5.48
171	Spinning, weaving and finishing of textiles	(C) Textiles	7.58	12.46
172	Manufacture of other textiles	e. Other misc. textiles	0.63	1.47
173	Manufacture of knitted and crocheted fabrics and articles	(C) Textiles	0.80	1.33

(continued)

A. Mitra, *Industry-Led Growth*, SpringerBriefs in Economics, DOI 10.1007/978-981-10-0009-6

Table A.1 (continued)

ASI code	Industry name	WPI used (1993–94 Base) to deflate value added	Average share in total value added (%)	Average share in total employment (%)
181	Manufacture of wearing apparel, except fur apparel (this class includes manufacture of wearing apparel made of material not made in the same unit. Both regular and contract activities are included)	b. Man made textiles	2.97	4.68
182	Dressing and dyeing of fur; manufacture of articles of fur	b1. Man made fibre	0.01	0.02
191	Tanning and dressing of leather, manufacture of luggage handbags, saddlery and Harness	Hessain and sacking bags	0.21	0.62
192	Manufacture of footwear	(F) Leather and leather products	0.46	1.16
201	Saw milling and planing of wood	Timber planks	0.02	0.12
202	Manufacture of products of wood, cork, straw and plaiting materials	b. Manufacture of board	0.21	0.50
210	Manufacture of paper and paper product paperboard	a. Paper and Pulp	1.71	2.13
221	Publishing (This group includes publishing whether or not connected with printing. Publishing involves financial, technical, artistic, legal and marketing activities, among others but not predominantly)	c. Printing and publishing of newspapers, periodicals etc.	0.64	0.63
222	Printing and service activities related to printing	c. Printing and publishing of newspapers, periodicals etc.	0.34	0.78
223	223 Reproduction of recorded media (This class includes reproduction of records, audio, video and computer tapes from master copies, reproduction of floppy, hard or compact disks, reproduction of non-customized software and film duplicating)	Roll films	0.07	0.03

(continued)

Table A.1 (continued)

ASI code	Industry name	WPI used (1993–94 Base) to deflate value added	Average share in total value added (%)	Average share in total employment (%)
231	Manufacture of coke oven products	III MANUFACTURED PRODUCTS	0.37	0.33
232	Manufacture of refined petroleum products	III MANUFACTURED PRODUCTS	8.63	0.58
241	Manufacture of basic chemicals	a. Basic heavy inorganic chemical	8.82	2.57
242	Manufacture of other chemical products	c1. Fertilizers	8.56	6.45
243	Manufacture of man-made fibres (This class includes manufacture of artificial or synthetic filament and non-filament fibers)	Man made fibre	0.47	0.31
251	Manufacture of rubber products	(G) Rubber and plastic products	1.63	1.37
252	Manufacture of plastic products	b. Plastic products	2.05	2.04
261	Manufacture of glass and glass products	b. Glass Earthernware Chinaware and their products	0.51	0.61
269	Manufacture of non-metallic mineral products n.e.c.	(I) Non-metallic mineral products	4.61	5.49
271	Manufacture of basic iron and steel	a1. Iron and steel	8.69	4.87
272	Manufacture of basic precious and non-ferrous metals	a2. Foundries for casting forging and structurals	2.28	0.95
273	Casting of metals (This group includes casting finished or semi-finished products producing a variety of goods, all characteristic of other activity classes)	a2. Foundries for casting forging and structurals	0.64	1.36
281	Manufacture of structural metal products, tanks, reservoirs and steam generators	c. Metal products	1.29	1.39

(continued)

Table A.1 (continued)

ASI code	Industry name	WPI used (1993–94 Base) to deflate value added	Average share in total value added (%)	Average share in total employment (%)
289	Manufacture of other fabricated metal products; metal working service activities	c. Metal products	1.87	2.42
291	Manufacture of general purpose machinery	a. Non-electrical machinery and parts	3.11	2.79
292	Manufacture of special purpose machinery	b. Electrical machinery	4.17	2.41
293	Manufacture of domestic appliances, n.e.c.	b4. Electrical apparatus and appliances	0.77	0.41
300	Manufacture of office, accounting and computing machinery	Computer and computer based system	2.0	2.0
311	Manufacture of electric motors, generators and transformers	Electrical generators	1.85	0.95
312	Manufacture of electricity distribution and control apparatus (electrical apparatus for switching or protecting electrical circuits (e.g. switches, fuses, voltage limiters, surge suppressors, junction boxes etc.) for a voltage exceeding 1000 V; similar apparatus (including relays, sockets etc.) for a voltage not exceeding 1000 V; boards, panels, consoles, cabinets and other bases equipped with two or more of the above apparatus for electricity control or distribution of electricity including power capacitors)	Power capacitors	1.39	0.69
313	Manufacture of insulated wire and cable (insulated (including enamelled or anodized) wire, cable (including coaxial cable) and other insulated conductors; insulated strip as is used in large capacity machines or control equipment; and optical fibre cables)	b2. Wires and cables	1.28	0.46

(continued)

Table A.1 (continued)

ASI code	Industry name	WPI used (1993–94 Base) to deflate value added	Average share in total value added (%)	Average share in total employment (%)
314	Manufacture of accumulators, primary cells and primary batteries	b3. Dry and wet batteries	2.06	0.24
315	Manufacture of electric lamps and lighting equipment	Gls lamps	0.74	0.29
319	Manufacture of other electrical equipment n.e.c.	Other electrical equipments and systems	0.29	0.32
321	Manufacture of electronic valves and tubes and other electronic components	Valve all types	0.74	0.000008
322	Manufacture of television and radio transmitters and apparatus for line telephony and line telegraphy	Telephone instruments	1.36	0.35
323	Manufacture of television and radio receivers, sound or video recording or reproducing apparatus, and associated goods	TV sets colour	2.0	0.37
331	Manufacture of medical appliances and instruments and appliances for measuring, checking, testing, navigating and other purposes except optical instruments	Medical X-ray films	1.21	0.57
332	Manufacture of optical instruments and photographic equipment	Roll films	0.19	0.07
333	Manufacture of watches and clocks	(K) Machinery and machine tools	0.19	0.18
341	Manufacture of motor vehicles	Motorcycles	3.32	0.96
342	Manufacture of bodies (coach work) for motor vehicles; manufacture of trailers and semi-trailers	Body manufactured for buses	0.09	0.25
343	Manufacture of parts and accessories for motor vehicles and their engines (brakes, gear boxes, axles, road wheels, suspension shock absorbers, radiators, silencers, exhaust pipes, steering wheels, steering columns and steering boxes and other parts and accessories n.e.c.)	Other automobile spare parts	2.70	2.46

(continued)

Table A.1 (continued)

ASI code	Industry name	WPI used (1993–94 Base) to deflate value added	Average share in total value added (%)	Average share in total employment (%)
351	Building and repair of ships and boats	a1. Heavy machinery and parts	0.17	0.28
352	Manufacture of railway and tramway locomotives and rolling stock	III MANUFACTURED PRODUCTS	0.21	0.30
353	Manufacture of aircraft and spacecraft	III MANUFACTURED PRODUCTS	0.10	0.05
359	Manufacture of transport equipment n.e.c.	b. Motor vehicles, motorcycles, scooters, bicycles and parts	2.22	1.70
361	Manufacture of furniture	Steel furnitures	0.24	0.33
369	Manufacturing n.e.c.	III MANUFACTURED PRODUCTS	1.13	1.46
371	Recycling of metal waste and scrap (from rejected aluminium, utensil, containers and other used metallic items etc. Collection of metal waste and scrap for recycling is included in 51,498)	(J) Basic metals alloys and metals products	0.003	0.007
372	Recycling of non-metal waste and scrap (from old new papers, rejected glass articles and used non-metallic items etc. Collection of non-metal waste and scrap for recycling is included in 51,498)	(J) Basic metals alloys and metals products	0.003	0.01
Other		MANUFACTURED PRODUCTS	1.92	1.72
Total		MANUFACTURED PRODUCTS		
Fixed asset		Machinery and machine tools		
Wages and salaries		Consumer price index for industrial workers		

Source Annual survey of industries, EPW research foundation

Table A.2 Employment growth in industries with a value added growth rate of more than 5 % per annum (ascending order)

Indus. code	Workers	Employees other than workers	Total persons	Value added
222	2.82	3.19	2.91	5.46
191	4.36	3.70	4.23	5.58
321	3.51	1.99	3.06	5.72
153	1.37	2.87	1.71	5.96
293	−1.32	−6.27	−2.60	6.49
221	0.46	3.29	1.92	6.65
252	6.01	1.24	4.58	6.73
192	6.76	2.55	6.14	6.75
223	1.64	−1.42	0.56	7.03
210	2.95	0.56	2.47	7.15
261	−0.45	−1.01	−0.55	7.77
155	5.29	2.46	4.55	7.83
242	2.75	2.77	2.76	8.10
351	2.90	−3.25	1.57	8.11
361	4.98	3.11	4.48	8.20
181	9.43	9.60	9.46	8.51
369	8.66	9.23	8.77	8.89
202	3.77	3.95	3.80	9.13
342	4.20	−2.27	2.99	9.18
322	−0.80	−6.63	−2.90	9.44
Total	2.98	1.29	2.58	9.45
272	0.99	−0.26	0.63	9.76
323	−4.05	−6.14	−4.70	9.84
271	2.91	2.22	2.74	10.05
291	0.73	−6.81	−2.61	10.07
172	9.12	19.46	13.48	10.11
269	5.85	3.81	5.44	10.40
289	7.63	4.08	6.78	10.62
292	3.36	0.01	2.26	10.80
315	3.08	0.72	2.57	11.50
182	4.42	8.40	4.91	11.55
343	8.70	5.36	7.88	11.63
319	8.66	3.25	7.04	12.32
371	8.81	8.88	9.31	12.72
332	4.92	5.00	5.01	13.68
314	−0.02	1.91	0.53	13.73
359	0.31	−2.15	−0.24	14.07
331	4.80	0.18	2.92	14.47
173	17.16	13.72	9.46	15.30

(continued)

Table A.2 (continued)

Indus. code	Workers	Employees other than workers	Total persons	Value added
281	7.00	3.65	6.13	16.51
311	2.78	0.16	1.93	17.07
312	7.04	2.18	5.50	18.22
231	3.45	1.06	2.95	18.43
Others	7.25	8.74	7.78	18.57
341	3.04	−1.65	1.69	18.97
300	8.10	3.97	6.06	22.88
232	6.60	0.10	4.71	25.46
372	43.15	44.12	44.26	59.44

Source See Table A.1

Table A.3 Industries with negative growth in total employment

Indus. code	Workers	Employees other than workers	Total persons	Value added
333	−6.08	−12.30	−7.55	−0.57
323	−4.05	−6.14	−4.70	9.84
243	−4.04	−6.35	−4.53	−14.00
322	−0.80	−6.63	−2.90	9.44
352	−2.21	−4.85	−2.86	4.92
291	0.73	−6.81	−2.61	10.07
293	−1.32	−6.27	−2.60	6.49
241	−1.68	−3.35	−2.24	−0.12
313	−0.09	−4.04	−1.21	−0.98
171	−0.80	−2.02	−0.98	4.62
160	−0.88	−0.91	−0.88	2.49
261	−0.45	−1.01	−0.55	7.77
359	0.31	−2.15	−0.24	14.07
201	−1.15	3.57	−0.04	4.59

Table A.4 Sluggishly growing industries in terms of total employment (0–2 % per annum)

Indus. code	Workers	Employees other than workers	Total persons	Value added
154	0.47	−1.22	0.13	2.66
314	−0.02	1.91	0.53	13.73
223	1.64	−1.42	0.56	7.03
272	0.99	−0.26	0.63	9.76
251	1.29	−0.53	0.87	2.29
351	2.90	−3.25	1.57	8.11
341	3.04	−1.65	1.69	18.97
153	1.37	2.87	1.71	5.96
152	3.34	−1.14	1.81	0.94
221	0.46	3.29	1.92	6.65
311	2.78	0.16	1.93	17.07

Table A.5 Industries with total employment growth of 2–4 % per annum

Indus. code	Workers	Employees other than workers	Total persons	Value added
Total	2.98	1.29	2.58	9.45
292	3.36	0.01	2.26	10.80
353	4.30	−0.36	2.37	3.59
151	3.12	0.40	2.44	0.32
210	2.95	0.56	2.47	7.15
315	3.08	0.72	2.57	11.50
271	2.91	2.22	2.74	10.05
242	2.75	2.77	2.76	8.10
222	2.82	3.19	2.91	5.46
331	4.80	0.18	2.92	14.47
231	3.45	1.06	2.95	18.43
342	4.20	−2.27	2.99	9.18
321	3.51	1.99	3.06	5.72
202	3.77	3.95	3.80	9.13
273	4.91	1.18	4.08	4.57

Table A.6 Rapidly growing industries in terms of total employment (in ascending order)

Indus. code	Workers	Employees other than workers	Total persons	Value added
191	4.36	3.70	4.23	5.58
361	4.98	3.11	4.48	8.20
155	5.29	2.46	4.55	7.83
252	6.01	1.24	4.58	6.73
232	6.60	0.10	4.71	25.46
182	4.42	8.40	4.91	11.55
332	4.92	5.00	5.01	13.68
269	5.85	3.81	5.44	10.40
312	7.04	2.18	5.50	18.22
300	8.10	3.97	6.06	22.88
281	7.00	3.65	6.13	16.51
192	6.76	2.55	6.14	6.75
289	7.63	4.08	6.78	10.62
319	8.66	3.25	7.04	12.32
Others	7.25	8.74	7.78	18.57
343	8.70	5.36	7.88	11.63
369	8.66	9.23	8.77	8.89
371	8.81	8.88	9.31	12.72
181	9.43	9.60	9.46	8.51
173	17.16	13.72	9.46	15.30
172	9.12	19.46	13.48	10.11
372	43.15	44.12	44.26	59.44

Table A.7 Employment elasticity with respect to value added and remuneration

Industry	Dep Var: Ln (person engaged)					Ln (workers)				
	Adj. R²	Ln (gross value added)		Ln (emoluments per all categories of employees including workers)		Adj. R²	Ln (gross value added)		Ln (wage per worker)	
		Coefficient	T	Coefficient	T		Coefficient	T	Coefficient	T
151	0.25	0.30	1.98*	0.26	0.82	0.03	0.25	1.45	0.20	0.22
152	0.49	-0.13	-1.03	0.60	3.25*	-0.18	0.33	0.69	-2.30	-0.78
153	0.90	0.29	2.71*	-0.02	-0.09	0.85	0.27	4.46*	-0.09	-0.38
154	0.09	0.10	1.26	0.01	0.06	0.40	0.13	2.57*	0.32	1.37
155	0.81	0.08	1.16	0.77	4.27*	0.63	0.24	2.49*	1.58	2.55*
160	0.26	0.28	1.18	-0.36	-2.17*	-0.22	-0.10	-0.54	-0.11	-0.16
171	0.38	0.39	2.59*	-0.79	-2.54*	-0.06	0.23	1.19	0.58	0.71
172	0.92	1.29	9.24*	-0.84	-5.02*	0.88	0.77	3.84*	0.17	0.22
173	0.95	1.12	9.71*	-0.92	-12.02	0.90	1.08	8.99*	-1.56	-1.86
181	0.94	0.40	2.48*	1.02	3.68	0.89	0.62	3.62*	1.08	2.16*
182	0.69	0.68	4.65*	-0.29	-0.76	0.70	0.70	4.74*	-0.59	-1.22
191	0.39	0.30	0.89	0.32	0.56	0.45	0.97	2.48*	-2.61	-1.50
192	0.89	0.66	4.51*	0.46	1.17	0.87	0.76	4.80*	1.00	1.01
201	0.05	0.08	1.50	-0.01	-0.11	0.00	0.10	1.04	-0.41	-1.39
202	0.62	0.10	0.57	0.52	1.41	0.48	0.31	2.13*	0.11	0.18
210	0.69	0.49	4.64*	-0.52	-2.44*	0.96	0.19	4.21*	-1.19	-9.68*
221	0.51	0.29	2.71*	-0.02	-0.14	0.55	0.20	2.90*	-0.63	-2.23*
222	0.78	0.33	3.60*	0.24	1.24	0.77	0.43	5.60*	0.10	0.20
223	0.61	0.64	3.35*	-0.69	-1.41	0.71	1.14	4.06*	1.38	1.51

(continued)

Table A.7 (continued)

Industry	Dep Var: Ln (person engaged)					Ln (workers)					
	Adj. R²	Ln (gross value added)		Ln (emoluments per all categories of employees including workers)		Adj. R²	Ln (gross value added)		Ln (wage per worker)	
		Coefficient	T	Coefficient	T		Coefficient	T	Coefficient	T
231	0.56	0.17	3.46*	-0.05	-0.18	0.67	0.15	2.67*	-0.31	-1.07
232	0.64	0.25	3.88*	-0.36	-1.55	0.79	0.25	5.63*	-0.56	-2.37*
241	0.69	0.42	2.44*	-0.36	-4.16*	0.28	0.36	1.59	-0.73	-1.68
242	0.89	0.30	2.80*	0.07	0.45	0.85	0.35	5.43*	-0.16	-0.11
243	0.54	0.10	0.63	-0.56	-1.60	0.67	0.20	2.55*	-1.30	-2.80*
251	0.26	0.39	1.97*	0.02	0.12	0.39	0.46	2.71*	-0.24	-0.80
252	0.73	0.50	1.26	0.27	0.42	0.88	0.85	6.33*	-0.31	-0.28
261	-0.20	-0.16	-0.43	0.16	0.27	-0.20	-0.02	-0.16	-0.17	-0.45
269	0.77	0.74	4.94*	-1.18	-2.50*	0.89	0.36	5.44*	-2.05	-5.00*
271	0.54	0.47	3.06*	-0.41	-1.34	0.76	0.26	3.26*	-0.90	-3.14*
272	0.49	0.35	3.11*	-0.30	-1.99*	0.59	0.28	3.80*	-0.35	-1.97*
273	0.90	0.66	7.07*	0.30	1.48	0.84	0.84	5.89*	0.22	0.32
281	0.85	0.47	5.70*	-0.22	-0.60	0.88	0.46	7.17*	-0.56	-1.18
289	0.97	0.60	8.17*	0.11	0.55	0.97	0.70	15.78*	0.14	0.26
291	0.63	0.35	2.33*	-0.95	-4.16*	0.31	0.06	0.53	-3.90	-2.18*
292	0.25	-0.04	-0.34	0.61	1.72	0.06	0.16	1.27	0.13	0.04
293	0.37	0.13	1.09	-1.26	-2.66*	-0.16	0.12	0.48	-0.06	-0.10
300	0.70	0.31	2.54*	-0.33	-0.72	0.99	0.27	2.32*	0.93	4.97*
311	0.27	0.28	1.89	-0.35	-0.72	0.42	0.16	1.11	-0.55	-0.71

(continued)

Table A.7 (continued)

Industry	Dep Var: Ln (person engaged)					Ln (workers)				
	Adj. R²	Ln (gross value added)		Ln (emoluments per all categories of employees including workers)		Adj. R²	Ln (gross value added)		Ln (wage per worker)	
		Coefficient	T	Coefficient	T		Coefficient	T	Coefficient	T
312	0.78	0.35	4.65*	−0.16	−0.44	0.82	0.38	2.76*	−0.23	−0.25
313	−0.26	−0.01	−0.18	0.12	0.35	−0.25	−0.01	−0.28	0.25	0.34
314	−0.27	0.00	−0.12	0.03	0.18	−0.04	−0.01	−0.34	−0.50	−1.25
315	−0.02	−0.05	−0.72	0.94	1.34	0.25	−0.02	−0.38	−1.18	−2.23*
319	0.72	−0.01	−0.06	1.14	2.53*	0.56	0.38	3.04*	0.88	0.80
321	0.05	0.02	0.07	0.30	0.89	−0.03	0.52	1.23	−0.82	−0.72
322	0.30	0.22	1.69	−1.07	−2.32*	0.63	−0.03	−0.24	−2.05	−3.67*
323	0.73	0.30	2.30*	−0.61	−4.79*	0.52	0.17	1.18	−1.60	−3.35*
331	0.65	0.86	3.80*	−1.26	−2.93*	0.70	0.33	4.05*	0.13	0.28
332	0.57	0.23	2.02*	0.10	0.31	0.57	0.27	2.85*	−0.15	−0.30
333	0.39	0.26	1.63	−0.99	−1.97	0.16	0.33	1.90	0.17	0.32
341	0.32	0.43	2.23*	−0.89	−1.70	0.84	0.20	5.59*	−1.60	−4.90*
342	0.72	0.56	3.33*	−0.33	−0.57	0.78	0.55	5.46*	−0.35	−0.80
343	0.81	0.39	1.42	0.69	0.98	0.84	0.57	4.45*	−2.39	−1.78
351	0.06	0.25	1.58	−0.38	−0.83	0.49	0.33	2.93*	−1.50	−2.33*
352	0.01	0.30	1.00	−0.80	−1.45	−0.17	0.29	0.81	0.88	0.73
353	0.44	0.24	2.47*	−0.02	−0.05	0.07	0.17	1.62	−0.07	−0.17
359	0.52	0.71	3.06*	−1.48	−3.40	0.89	0.49	6.39*	−2.68	−8.47*
361	0.68	0.17	1.15	0.53	2.04	0.55	0.36	2.64*	−0.74	−1.36

(continued)

Table A.7 (continued)

Industry	Dep Var: Ln (person engaged)					Ln (workers)					
	Adj. R²	Ln (gross value added)		Ln (emoluments per all categories of employees including workers)		Adj. R²	Ln (gross value added)		Ln (wage per worker)	
		Coefficient	T	Coefficient	T		Coefficient	T	Coefficient	T
369	0.96	0.36	3.98*	1.05	5.89	0.95	0.53	6.18*	1.25	4.45*
371	0.96	0.61	2.60*	-0.41	-1.04	0.98	0.91	6.40*	-0.97	-3.94*
372	0.94	0.75	3.68*	-0.65	-1.73	0.95	0.75	5.38*	-0.69	-2.62*
Others	1.00	0.12	1.45	0.87	4.64	1.00	0.33	5.39*	0.35	2.41*
Total	0.80	0.43	2.79*	-0.21	-0.77	0.89	0.35	8.83*	-0.54	-0.85

Note *represents significance at 5 % level
Source Based on ASI data (1998–99 through 2007–08)

Table A.8 Decomposition of value added in terms of productivity and employment

Ind. code	Value added per person engaged	Persons engaged
151	−2.12	2.44
152	−0.87	1.81
153	4.25	1.71
154	2.53	0.13
155	3.28	4.55
160	3.37	−0.88
171	5.60	−0.98
172	−3.37	13.48
173	5.84	9.46
181	−0.95	9.46
182	6.64	4.91
191	1.36	4.23
192	0.61	6.14
201	4.63	−0.04
202	5.33	3.80
210	4.67	2.47
221	4.73	1.92
222	2.55	2.91
223	6.47	0.56
231	15.49	2.95
232	20.75	4.71
241	2.12	−2.24
242	5.34	2.76
243	−9.47	−4.53
251	1.42	0.87
252	2.16	4.58
261	8.33	−0.55
269	4.95	5.44
271	7.31	2.74
272	9.13	0.63
273	0.49	4.08
281	10.38	6.13
289	3.84	6.78
291	12.68	−2.61
292	8.53	2.26
293	9.09	−2.60
300	16.82	6.06
311	15.14	1.93
312	12.72	5.50
313	0.24	−1.21

(continued)

Table A.8 (continued)

Ind. code	Value added per person engaged	Persons engaged
314	13.20	0.53
315	8.92	2.57
319	5.28	7.04
321	2.65	3.06
322	12.34	−2.90
323	14.54	−4.70
331	11.55	2.92
332	8.67	5.01
333	6.98	−7.55
341	17.29	1.69
342	6.19	2.99
343	3.75	7.88
351	6.54	1.57
352	7.78	−2.86
353	1.22	2.37
359	14.30	−0.24
361	3.73	4.48
369	0.12	8.77
371	3.41	9.31
372	15.18	44.26
Others	10.80	7.78
Total	6.87	2.58

Table A.9 Wages and salaries

Ind. code	Wage share	Salary share	Emolument share	Wage rate/salary per employee	Worker/employee
151	11.51	10.38	21.88	0.37	3.18
152	15.31	20.17	35.48	0.40	1.98
153	14.92	9.91	24.83	0.45	3.37
154	18.42	16.30	34.72	0.30	3.94
155	10.06	11.25	21.32	0.31	3.01
160	16.68	6.23	22.90	0.17	18.14
171	16.92	8.95	25.87	0.34	5.92
172	15.02	11.93	26.95	0.51	3.86
173	13.11	8.44	21.55	0.20	4.01
181	11.51	7.99	19.50	0.26	5.97
182	15.34	10.67	26.00	0.30	5.80
191	20.44	16.09	36.53	0.31	4.29
192	18.53	12.28	30.81	0.29	5.54
201	24.76	13.63	38.40	0.63	3.05
202	14.79	11.75	26.54	0.38	3.37
210	13.38	11.12	24.50	0.34	3.69
221	11.31	28.47	39.78	0.44	0.94
222	20.42	19.14	39.56	0.37	3.02
223	3.58	8.05	11.63	0.30	1.63
231	18.13	10.44	28.56	0.44	3.83
232	2.72	3.01	5.73	0.36	2.58
241	5.02	7.57	12.59	0.36	1.97
242	6.34	11.91	18.25	0.26	2.10
243	19.66	17.99	37.65	0.35	3.78
251	10.47	9.39	19.85	0.36	3.29
252	7.17	8.98	16.15	0.30	2.83
261	12.93	10.78	23.71	0.30	4.15
269	9.24	7.98	17.22	0.28	4.26
271	10.75	10.13	20.89	0.37	3.01
272	7.71	7.03	14.74	0.40	2.84
273	21.01	18.18	39.19	0.36	3.43
281	12.69	14.22	26.91	0.33	2.78
289	11.86	11.47	23.33	0.33	3.27
291	11.72	16.73	28.45	0.49	1.73
292	8.46	12.42	20.88	0.34	2.06
293	6.29	6.66	12.95	0.34	2.82
300	1.49	3.84	5.33	0.26	1.53
311	9.50	12.28	21.78	0.38	2.03
312	6.65	9.94	16.59	0.32	2.11

(continued)

Table A.9 (continued)

Ind. code	Wage share	Salary share	Emolument share	Wage rate/salary per employee	Worker/employee
313	7.02	8.53	15.54	0.34	2.59
314	7.37	7.88	15.25	0.37	2.49
315	8.94	8.22	17.16	0.32	3.48
319	11.22	14.35	25.57	0.31	2.70
321	9.82	16.92	26.74	0.29	2.16
322	3.94	7.32	11.25	0.32	1.75
323	1.92	4.21	6.13	0.23	2.11
331	5.17	9.83	15.00	0.33	1.71
332	3.77	5.50	9.27	0.30	2.33
333	14.95	15.28	30.23	0.31	3.42
341	8.95	9.91	18.86	0.36	2.51
342	29.04	17.27	46.31	0.39	4.59
343	11.95	12.93	24.88	0.32	3.05
351	27.25	18.85	46.10	0.42	3.60
352	18.03	16.10	34.13	0.39	2.98
353	21.61	28.42	50.02	0.50	1.49
359	10.77	10.53	21.30	0.30	3.42
361	17.31	16.74	34.06	0.44	2.59
369	13.71	9.99	23.70	0.37	3.84
371	30.71	16.09	55.53	0.44	4.90
372	24.01	14.34	38.35	0.44	4.20
Others	7.12	9.05	16.17	0.41	1.90
Total	10.03	10.51	20.54	0.29	3.38

Table A.10 Inputs drawn from various sub-sectors for the production of 1 unit of output (Column total is 1)

	Food S14	Bev S15	Tex S16	Other tex S17	Wood S18	Paper S19	Leather S20	Rubber S21	Petroleum S22	Fertilizers S23	Chemicals S24	Cement S25	Non-metals S26	Metals S27	Metal products S28	Non-electricals S29	Elect machinary S30	Transports S31	Other mfg S32
Food	0.015	0.000	0.000	0.000	0.000	0.003	0.892	0.000	0.026	0.000	0.000	0.000	0.001	0.595	0.456	0.002	0.000	0.003	0.005
Bev	0.000	0.000	0.000	0.000	0.000	0.000	0.000	0.000	0.000	0.000	0.000	0.000	0.001	0.004	0.193	0.000	0.000	0.008	0.000
Tex	0.011	0.002	0.000	0.000	0.000	0.006	0.005	0.001	0.109	0.000	0.000	0.000	0.016	0.005	0.008	0.492	0.567	0.023	0.004
Other tex	0.000	0.000	0.000	0.002	0.000	0.000	0.000	0.043	0.209	0.000	0.000	0.000	0.001	0.008	0.012	0.023	0.082	0.010	0.005
Wood	0.000	0.000	0.000	0.000	0.001	0.000	0.001	0.003	0.016	0.043	0.004	0.007	0.009	0.035	0.007	0.007	0.010	0.091	0.040
Paper	0.002	0.001	0.000	0.000	0.000	0.001	0.003	0.038	0.000	0.009	0.001	0.005	0.007	0.071	0.053	0.021	0.014	0.081	0.641
Leather	0.000	0.000	0.000	0.002	0.000	0.000	0.004	0.000	0.000	0.000	0.000	0.000	0.000	0.000	0.000	0.001	0.015	0.003	0.000
Rubber	0.000	0.000	0.000	0.000	0.000	0.000	0.004	0.096	0.000	0.013	0.001	0.084	0.025	0.037	0.082	0.034	0.031	0.114	0.029
Petroleum	0.189	0.192	0.068	0.161	0.241	0.145	0.005	0.396	0.361	0.060	0.179	0.307	0.139	0.075	0.028	0.067	0.029	0.036	0.041
Fertilizers	0.651	0.696	0.349	0.804	0.738	0.711	0.000	0.003	0.000	0.000	0.000	0.000	0.000	0.011	0.000	0.000	0.000	0.002	0.000
Chemicals	0.081	0.092	0.576	0.012	0.008	0.111	0.051	0.002	0.023	0.420	0.159	0.237	0.344	0.099	0.134	0.298	0.142	0.238	0.185
Cement	0.000	0.000	0.000	0.000	0.000	0.000	0.000	0.000	0.000	0.000	0.002	0.000	0.004	0.000	0.000	0.000	0.000	0.000	0.000
Non-metals	0.000	0.000	0.000	0.000	0.000	0.000	0.001	0.000	0.000	0.000	0.099	0.000	0.034	0.000	0.002	0.000	0.001	0.005	0.003
Metals	0.000	0.000	0.000	0.000	0.000	0.000	0.002	0.002	0.016	0.000	0.002	0.002	0.244	0.001	0.000	0.001	0.003	0.153	0.010
Metal products	0.000	0.000	0.000	0.000	0.000	0.000	0.002	0.020	0.016	0.070	0.184	0.289	0.042	0.004	0.003	0.001	0.005	0.072	0.006
Non-electricals	0.044	0.012	0.005	0.014	0.009	0.019	0.020	0.046	0.000	0.297	0.361	0.041	0.103	0.047	0.022	0.044	0.077	0.049	0.007
Elect machinary	0.000	0.000	0.000	0.000	0.000	0.000	0.003	0.025	0.000	0.000	0.000	0.013	0.004	0.003	0.000	0.001	0.004	0.073	0.012
Transports	0.004	0.003	0.001	0.004	0.002	0.003	0.000	0.193	0.224	0.056	0.005	0.008	0.007	0.001	0.000	0.000	0.001	0.017	0.000
Other mfg	0.001	0.000	0.000	0.001	0.000	0.001	0.005	0.131	0.000	0.031	0.004	0.007	0.018	0.004	0.000	0.007	0.021	0.022	0.013

Source Pradhan et al. (2013)

Table A.11 Distribution of 1 unit of output from a given sub-sector as input across various sub-sectors (row total is 1)

	Food S14	Bev S15	Tex S16	Other tex S17	Wood S18	Paper S19	Leather S20	Rubber S21	Petroleum S22	Fertilizers S23	Chemicals S24	Cement S25	Non-metals S26	Metals S27	Metal products S28	Non-electricals S29	Elect machinary S30	Transports S31	Other mfg S32
Food	0.610	0.296	0.002	0.000	0.000	0.003	0.000	0.002	0.000	0.001	0.078	0.000	0.001	0.001	0.003	0.001	0.001	0.000	0.002
Bev	0.026	0.849	0.001	0.000	0.007	0.000	0.000	0.005	0.000	0.001	0.098	0.000	0.002	0.004	0.005	0.000	0.001	0.000	0.001
Tex	0.005	0.005	0.395	0.513	0.003	0.002	0.003	0.010	0.000	0.000	0.024	0.001	0.004	0.002	0.002	0.003	0.017	0.005	0.006
Other tex	0.044	0.044	0.110	0.447	0.007	0.014	0.011	0.063	0.003	0.004	0.089	0.008	0.010	0.008	0.009	0.035	0.072	0.009	0.013
Wood	0.134	0.017	0.024	0.034	0.040	0.077	0.004	0.025	0.008	0.012	0.206	0.024	0.031	0.024	0.027	0.091	0.144	0.033	0.044
Paper	0.105	0.050	0.026	0.019	0.014	0.488	0.002	0.025	0.005	0.000	0.088	0.012	0.019	0.012	0.012	0.020	0.074	0.010	0.019
Leather	0.000	0.000	0.007	0.085	0.002	0.000	0.640	0.085	0.000	0.000	0.033	0.000	0.001	0.005	0.003	0.007	0.044	0.069	0.020
Rubber	0.036	0.051	0.028	0.028	0.013	0.015	0.004	0.222	0.006	0.004	0.131	0.028	0.018	0.020	0.023	0.045	0.200	0.079	0.049
Petroleum	0.053	0.012	0.039	0.019	0.003	0.015	0.003	0.029	0.163	0.103	0.166	0.020	0.081	0.152	0.029	0.026	0.058	0.020	0.010
Fertilizers	0.106	0.001	0.000	0.000	0.002	0.000	0.000	0.013	0.000	0.672	0.205	0.000	0.001	0.000	0.000	0.000	0.000	0.000	0.000
Chemicals	0.020	0.017	0.048	0.026	0.005	0.019	0.004	0.129	0.023	0.035	0.522	0.000	0.010	0.022	0.009	0.012	0.069	0.017	0.014
Cement	0.001	0.000	0.000	0.002	0.001	0.000	0.003	0.007	0.035	0.000	0.066	0.014	0.724	0.032	0.024	0.013	0.029	0.040	0.012
Non-metals	0.001	0.004	0.001	0.002	0.002	0.005	0.001	0.006	0.002	0.000	0.029	0.119	0.418	0.053	0.021	0.035	0.258	0.018	0.023
Metals	0.000	0.000	0.000	0.000	0.003	0.001	0.000	0.009	0.000	0.000	0.009	0.000	0.003	0.330	0.116	0.161	0.281	0.064	0.022
Metal products	0.003	0.001	0.001	0.004	0.006	0.002	0.001	0.015	0.000	0.000	0.008	0.000	0.007	0.184	0.081	0.163	0.359	0.129	0.035
Non-elec	0.035	0.010	0.027	0.053	0.004	0.003	0.003	0.015	0.006	0.001	0.039	0.001	0.006	0.037	0.042	0.294	0.244	0.144	0.034
Elect machinary	0.001	0.000	0.000	0.001	0.003	0.002	0.000	0.005	0.000	0.000	0.007	0.000	0.000	0.015	0.016	0.070	0.792	0.054	0.033
Transports	0.004	0.000	0.000	0.001	0.005	0.000	0.001	0.019	0.000	0.000	0.001	0.000	0.000	0.066	0.010	0.039	0.016	0.833	0.006
Other mfg	0.003	0.000	0.005	0.017	0.002	0.005	0.003	0.008	0.002	0.000	0.039	0.001	0.008	0.013	0.017	0.038	0.092	0.040	0.708

Reference

Pradhan BK, Saluja MR, Sharma AK (2013) Social accounting matrix for India 2007–08, IEG Working Paper (Delhi, Institute of Economic Growth)